BONDED STATIONARY PHASES IN CHROMATOGRAPHY

BONDED STATIONARY PHASES IN CHROMATOGRAPHY

Edited by

Eli Grushka, PhD
Department of Chemistry
State University of New York at Buffalo

ann arbor science PUBLISHERS INC.
POST OFFICE BOX 1425 • ANN ARBOR, MICHIGAN 48106

P.O. Box 1425, Ann Arbor, Michigan 48106

Library of Congress Catalog Card No. 73-86056
ISBN 0-250-40039-1

Manufactured in the United States of America

PREFACE

Bonded stationary phases are utilized now in many laboratories for practical applications, and their influence on the chromatographic processes is being studied intensely. It is well known that modification of solid support particles by means of chemical reactions with organic compounds can alter chromatographic behavior. Such modification has been utilized for surface deactivation of support in gas chromatography since the late 1950's. In 1969 Halasz and Sebastian, by describing the usefulness of bonded phases in their own right as packing for gas chromatographic systems, prompted the scientific community to examine carefully the impact of such phases on the more general separation problems. On the other hand, biochemists and biologists, realized in the early 1960's that by binding antigens, co-enzymes, etc. they could achieve specific separation of antibodies, enzymes, and other biologically active substances. Thus affinity chromatography was developed.

The past five years have seen a tremendous growth in the popularity and utilization of bonded phases. This is understandable since, among others, bonded phases in gas chromatography allow operation at higher temperature, while in liquid chromatography the requirement of presaturator column is eliminated. On the other hand, it is generally recognized that the same phase can behave differently (in a chromatographic sense) when it is bonded to the support than when coated conventionally. Despite numerous practical applications, many questions regarding the retention mechanism in bonded phase systems remain unanswered. It is not entirely clear whether the phase acts as a surface modifier, as an adsorbant in its own right or as its bulk liquid. The degree of cross-polymerization of some bonded phases can affect the retention; how and to what extent are not yet completely known. Bonded phases affect the efficiency of chromatographic columns but literature reports conflict as to the degree. The main purpose of this book is to summarize and indicate some of the advances made in the fascinating field of bonded phases in the hope that it will instigate the reader to consider such systems more systematically. We also hope that the book will enable the researcher to select the best system for his particular separation.

We do not wish to encourage proliferation of many bonded phases, each suited for a particular separation. Recent studies on gas chromatographic phases seem to indicate that many of these phases have similar separation abilities. Further work should be directed, perhaps, at a better understanding of bonded phase characteristics, thus developing a relatively small number of systems that could be utilized for multiseparation purposes.

E. Grushka
June 1974

TABLE OF CONTENTS

CONTRIBUTORS

Walter A. Aue, 5637 Life Sciences Building, Dalhousie University, Halifax, N.S., B3H 3J5, Canada

Richard L. Cotter, Waters Associates, Inc., Maple Street, Milford, Massachusetts

Augustus M. Filbert, Corning Glass Works, Sullivan Park, Science Center, Corning, New York

Dennis F. Horgan, Jr., Waters Associates, Inc., Maple Street, Milford, Massachusetts

Csaba Horvath, Department of Engineering and Applied Science, Yale University, New Haven, Connecticut

Shubhender Kapila, Environmental Trace Substances Research Center, University of Missouri, Columbia, Missouri

James N. Little, Waters Associates, Inc., Maple Street, Milford, Massachusetts

Merrill Lynn, Corning Glass Works, Sullivan Park, Science Center, Corning, New York

R. E. Majors, Varian, Instrument Division, 2700 Mitchell Drive, Walnut Creek, California

William E. Meyers, Department of Biochemistry, Ohio State University, 484 W. 12th Avenue, Columbus, Ohio

Milos Novotny, Department of Chemistry, Indiana University, Bloomington, Indiana

Wolfgang Parr, Department of Chemistry, University of Houston, Houston, Texas

Fredric M. Rabel, H. Reeve Angel & Co., R&D Laboratory, 568 Myrtle Avenue, Clifton, N.J.

Garfield P. Royer, Department of Biochemistry, Ohio State University, 484 W. 12th Avenue, Columbus, Ohio

Richard V. Vivilecchia, Waters Associates, Inc., Maple Street, Milford, Massachusetts

Reed C. Williams, Instrument Products Division, E. I. du Pont de Nemours & Co., Inc., Wilmington, Delaware

CHAPTER 1

CHEMICALLY BONDED PACKINGS FOR CHROMATOGRAPHY

Merrill Lynn and Augustus M. Filbert

INTRODUCTION

Packing materials with a liquid phase covalently bonded to solid, inorganic supports offer advantages over other commonly used packings for both liquid-liquid and gas-liquid chromatographic separations. The chemically bonded phases are more thermally stable and are more stable toward solvents than standard liquid phases.[1,2] Chemical coupling also reduces the surface activity of inorganic supports.

COVALENT BONDING TO SUPPORTS

One approach to developing chemically bonded liquid-phase packings is through the reaction of silanes with inorganic supports. The work described here was performed with octadecyltrichlorosilane and γ-aminopropyltriethoxysilane,[3,4] but a large variety of packings can be made by judicious selection of the silane (Table 1.1).

Packings ranging from nonpolar to highly polar can be made by direct reaction of the support with a silane possessing the desired functional groups. In addition, many more packings can be made by using organic chemical reactions to modify the silanized materials.

The work presented below was carried out with silica and controlled-pore glass support systems. Scanning electron micrographs of a controlled-pore

Table 1.1

Groups Available by Direct Reaction with Silanes

$-CH_3$	$-CH_2Cl$
$-(CH_2)_{17}-CH_3$	$-CH_2CN$
$-CH=CH_2$	$-CH_2NH_2$
$-C_6H_5$	$-CH_2SH$
$-C_6H_4-X$	$-O-CH_2-CH-CH_2$ (epoxide, CH and CH_2 bridged by O)

glass at various magnifications are shown in Figures 1.1, 1.2 and 1.3. Information on the preparation of controlled-pore glass has been discussed by Hair and Filbert,[5] as have the surface properties of this unique series of glasses.[6]

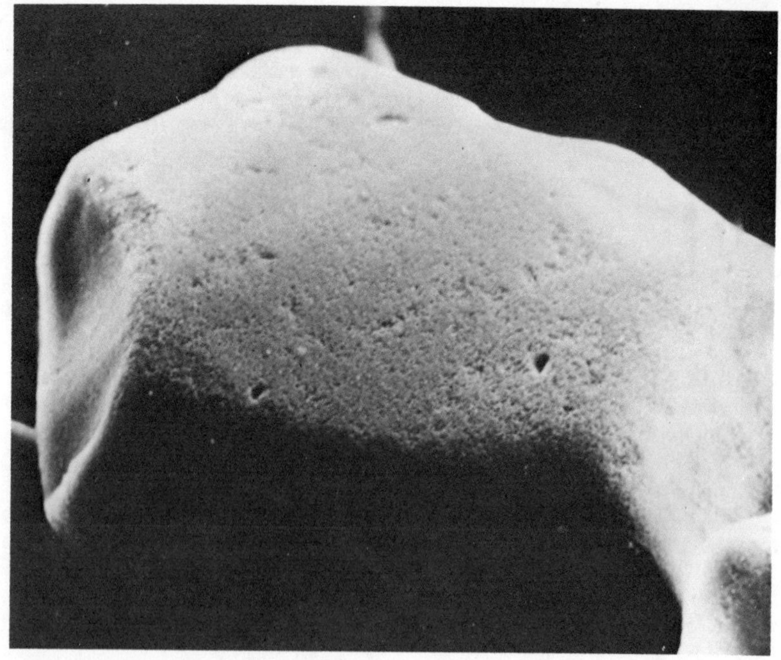

Figure 1.1. Scanning electron micrograph 315 X, 2300 Å controlled-pore glass. 40μm

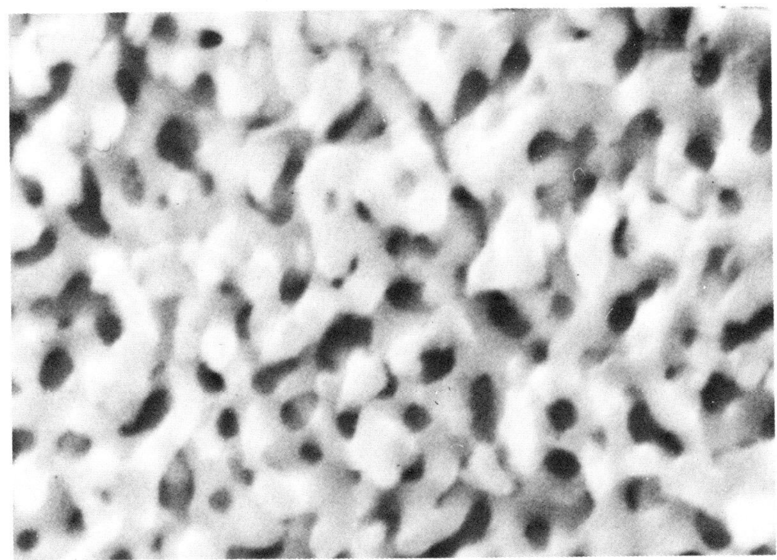

Figure 1.2. Scanning electron micrograph 10,500 X, 2300 Å controlled-pore glass. 1.2µm

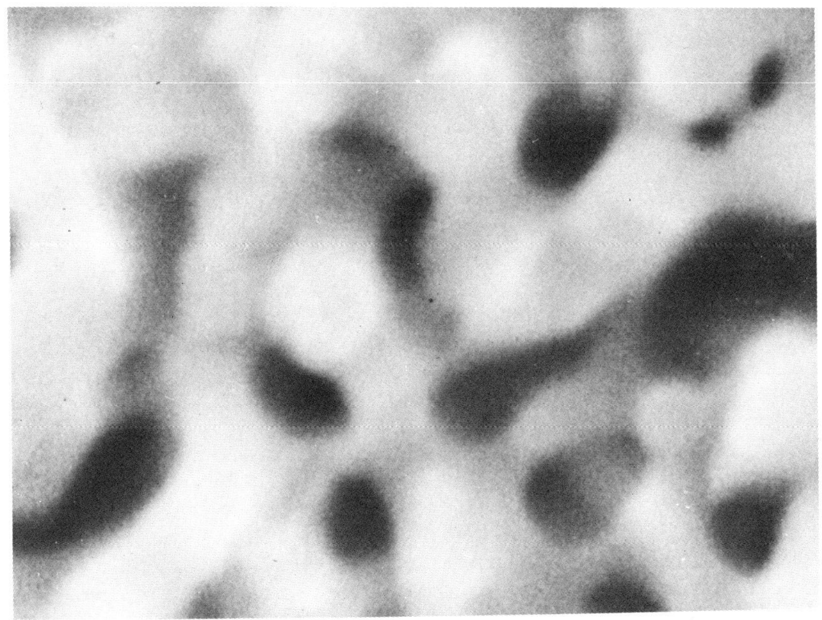

Figure 1.3. Scanning electron micrograph 31,500 X, 2300 Å controlled-pore glass. 0.4µm

The surface properties of both silica and controlled-pore glass are affected by heat treatment, as shown by the infrared spectra in Figure 1.4.

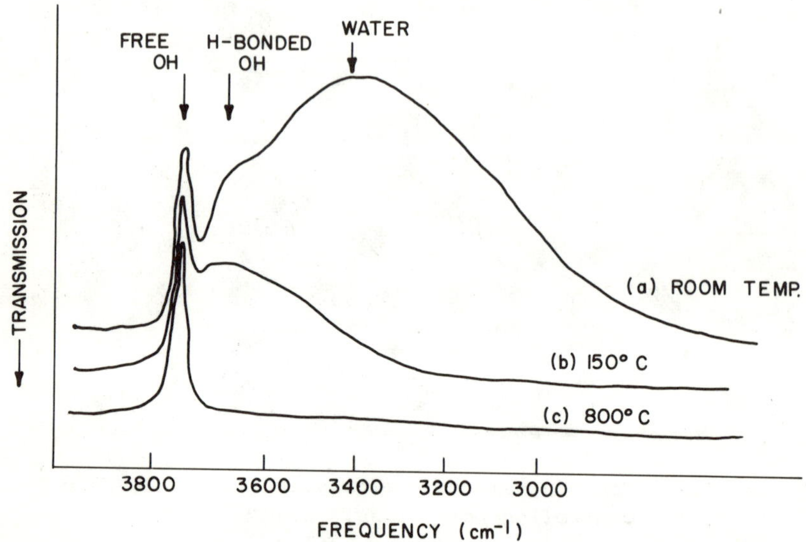

Figure 1.4. Infrared spectra of a high surface area silica.

Trace (a) is the spectra of a wet silica surface, trace (b) is the spectra of silica heated to 150°C, and trace (c) is the spectra of silica heated to 800°C. The broad band at 3450 cm^{-1} in trace (a) is associated with molecular water, which is removed when the support is heated to 150°C. The band at 3660 cm^{-1} is attributed to surface hydroxyl groups that are close together and are hydrogen-bonded to each other. As the silica is heated further, this band disappears, and at 500°C occurs only as a tail. At 800°C the 3660 cm^{-1} completely disappears, leaving only the band at 3747 cm^{-1}. The latter band is attributed to the vibrations of a free silanol group.

The mechanism of silane reaction with the silica support surface involves condensation of the ester, halide or silanol functional group on the silane with the silanols on the silica surface. In the absence of water on the silica surface, the halide or ester groups on the silane condense with the freely vibrating silanol groups on the silica.

A 290 Å controlled-pore glass disc that was first heated to 500°C and then reacted with silane is shown in Figure 1.5. Only the free silanol band at 3747 cm^{-1} and a small band at 3690 cm^{-1} attributed to boronols[7] are present. The silanol groups were reacted with octadecyltrichlorosilane as the first step in the preparation of a permanent liquid-phase packing.

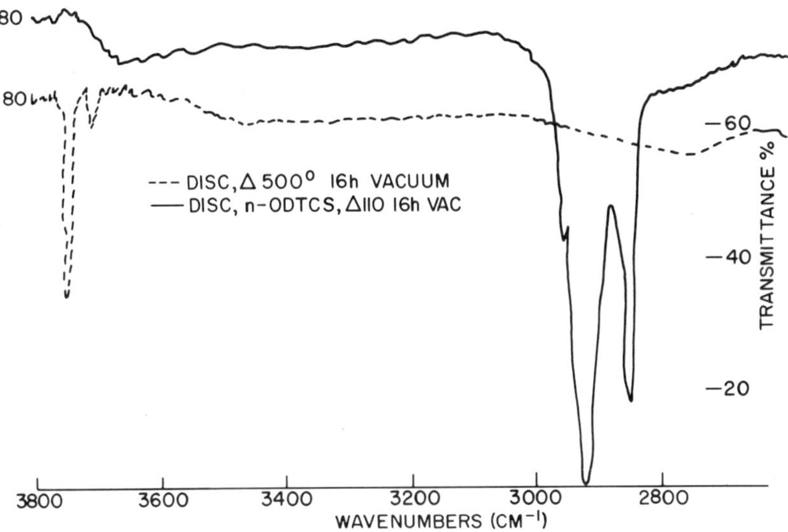

Figure 1.5. Infrared spectra of a controlled-pore glass disc treated with n-octadecyltrichlorosilane.

The support was refluxed in a toluene solution of the silane, and then refluxed in toluene to remove all the silane except those molecules that were chemically bonded to the surface. This technique should produce a liquid phase that is as close to a monolayer as can be achieved. The infrared spectra after reaction (Figure 1.5) shows the disappearance of the silanol and boronol peaks, and the appearance of bands attributed to the $-CH_2-$ and $-CH_3-$ groups on the octadecylsilane derivative.

In the presence of water the alkoxy groups on the silane hydrolyze to form hydroxyls which react with the silanols on the glass surface.

$$1.\quad (RO)_3\,Si\text{-}CH_2\text{-}CH_2\text{-}CH_2\text{-}NH_2 \xrightarrow{H_2O} (HO)_3Si\text{-}CH_2\text{-}CH_2\text{-}CH_2\text{-}NH_2$$

$$2.\quad \diagdown\!\!\!\!\diagup Si\text{-}OH + HO\text{-}Si(OH)_2\text{-}CH_2\text{-}CH_2\text{-}CH_2\text{-}NH_2 \longrightarrow \diagdown\!\!\!\!\diagup Si\text{-}O\text{-}Si(O\text{-})_2\text{-}CH_2\text{-}CH_2\text{-}CH_2\text{-}NH_2$$

Ideally, it would be desirable for all of the hydroxyl groups on the silane to react with surface silanols. This does not occur, however, because of steric factors, and instead a partial condensation between silanes occurs to form siloxane polymer. In addition, there are some unreacted hydroxyl groups, but under these conditions the liquid phase must be considered to be a multi-layer coating.

SILANIZATION STUDIES

Some of the factors which may affect the concentration or thickness of the organic layer are silane monomer concentration, reaction solvent, pH, reaction temperature, reaction time, and surface area. The concentration of reactive silanols on the silica surface is determined by the surface area and thermal history of the material. As described previously, heating the support to a high temperature and silanizing in a nonaqueous system should provide a monomolecular layer on the glass. In general, however, higher loadings of silane are obtained from organic solvents than aqueous systems if there is water on the surface of the glass during the reaction.

As is readily expected, silane loadings are affected by the concentration of monomeric silane in the reaction solution (Figure 1.6). At low silane concentrations small increases in concentration have a large effect on silane loading, but above the 5% level there is very little effect from increasing concentration. In this case the support is a 550 Å controlled-pore glass with a surface area of 70 m^2/g.

With the same glass and a 10% aqueous silane solution, the silane loadings increased with increasing temperature up to approximately 75°C (Figure 1.7). Silane loadings were also affected by pH changes in the range of pH 6 to 11 (Figure 1.8).

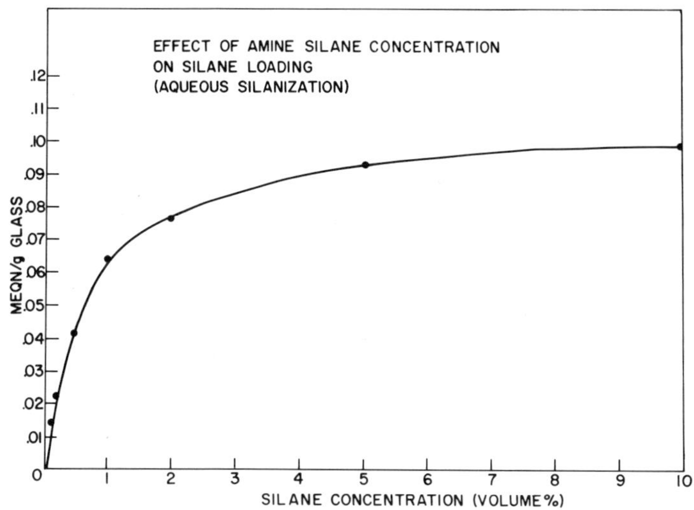

Figure 1.6. Effect of silane concentration on silane loading of controlled-pore glass (aqueous amine silane).

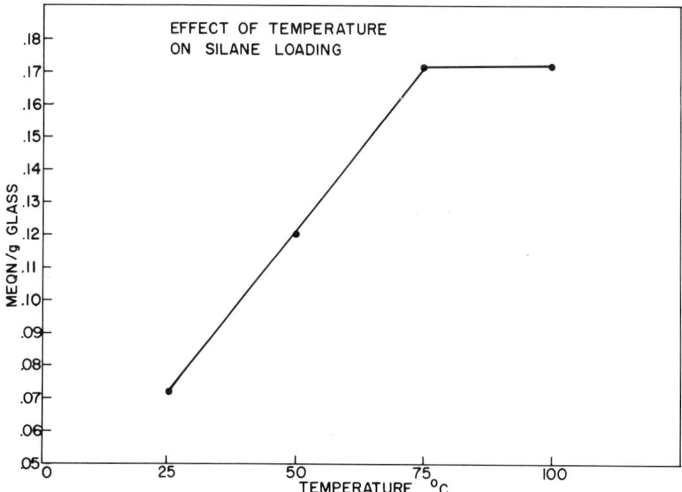

Figure 1.7. Effect of temperature on silane loading of controlled-pore glass.

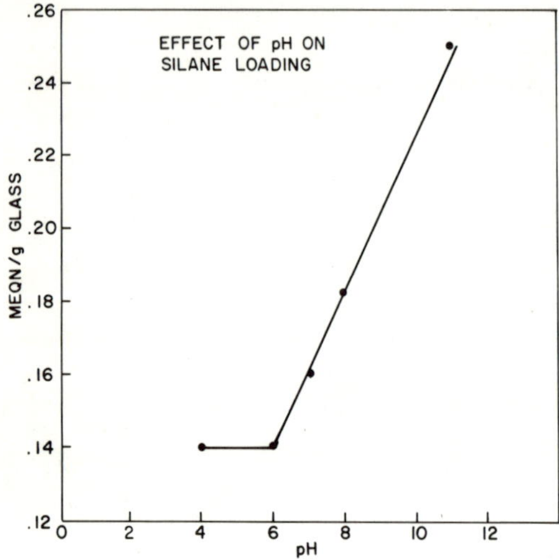

Figure 1.8. Effect of pH on silane loading of controlled-pore glass.

In this case the highest loading was produced at pH 11, which is the pH of a 10% solution of γ-aminopropyltriethoxy silane.

The above reactions were all run for two-hour time periods in order to eliminate effects of reaction time. However, reaction time does not appear to be a critical factor when the silanization reaction is run at elevated temperatures (Table 1.2).

Table 1.2

Effect of Reaction Time on Silane Loading (aqueous silanization)

Reaction Time (minutes)	*MEQN per g-Glass*
0	0.15
15	0.16
30	0.16
60	0.17
120	0.18
240	0.18

STABILITY OF PACKINGS

The permanent liquid-phase packings were evaluated for both thermal and chemical stability. The thermal stability of the γ-aminopropylsilane packing was examined with thermal gravimetric analysis and differential thermal analysis. With both thermal analyses the packing was stable up to 300°C. At higher temperatures there was a loss of silane from the carrier. Under similar circumstances a silicone oil-treated glass packing showed appreciable losses between 100 and 200°C.

An octadecylsilane-controlled-pore glass packing was evaluated with infrared studies. The effect of heat treatment on an octadecylsilane-porous glass disc is shown in Figure 1.9. As the temperature is increased to 300°C, there is no effect on the adsorption bands associated with carbon-hydrogen stretching. At higher temperatures, there is a loss of silane from the surface of the disc.

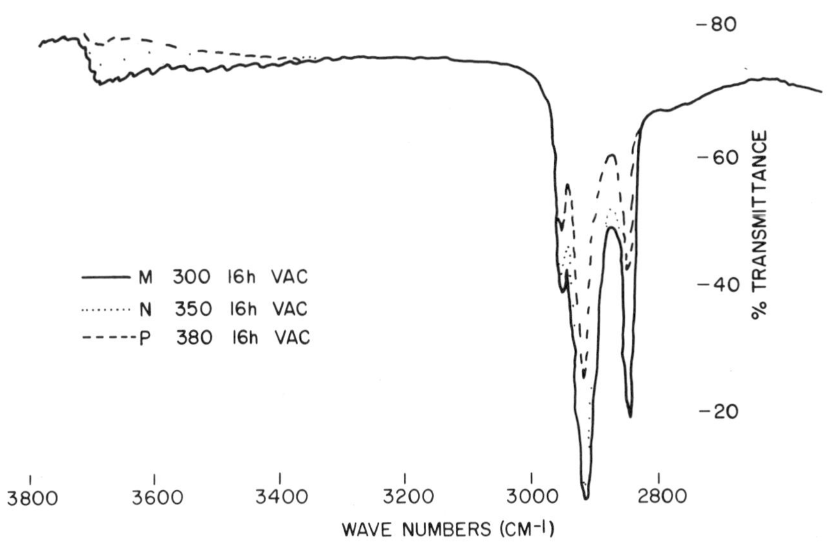

Figure 1.9. Infrared spectra of a heat-treated octadecylsilane-porous glass disc.

Chemical stability of the octadecylsilane packing was evaluated by refluxing 300 Å porous glass discs in a series of solvents and examining the infrared spectra of the discs before and after extraction (Figure 1.10). No changes in the intensity of the carbon-hydrogen stretching bands were observed, which indicates that silane had not been removed from the surface.

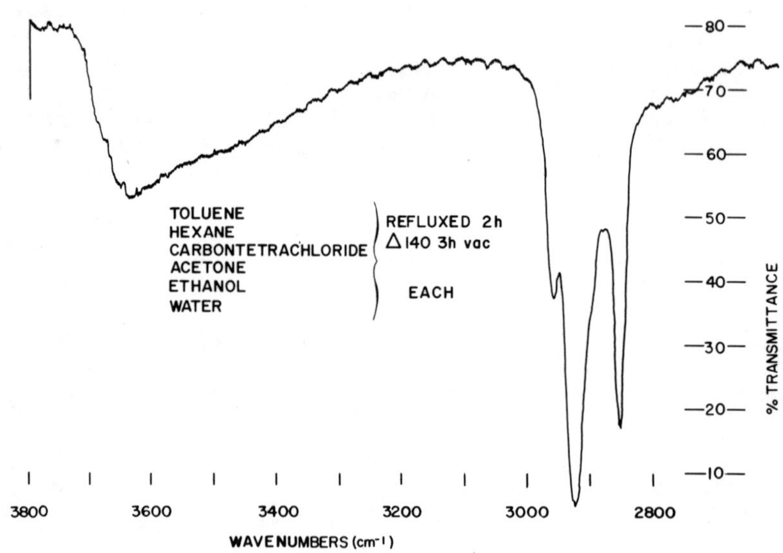

Figure 1.10. Infrared spectra of solvent treated octadecylsilane-porous glass discs.

SUMMARY

The silanization technique is an excellent method for preparing packing materials with chemically bonded liquid phases. The availability of a large number of silanes allows the preparation of a wide variety of liquid phases ranging from nonpolar to polar. Inorganic supports can be readily modified by silanization to provide almost any type of surface for gas-liquid or liquid-liquid chromatographic separations.

REFERENCES

1. Karasek, F. W. *Res. Devel., 22,* 34 (1971).
2. Majors, R. E. *Am. Lab., 4,* 27 (1972).
3. Sterman, S. and J. G. Marsden. Union Carbide Corp., Bulletin F-41920 (1968).
4. Aue, W. A. and C. R. Hastings. *J. Chromatog., 42,* 319 (1969).
5. Hair, M. L. and A. M. Filbert. *Res. Devel., 20(12),* 30 (1969).
6. Hair, M. L. and A. M. Filbert. *Res. Devel., 20(10),* 34 (1969).
7. Hair, M. L. and I. D. Chapman. *J. Am. Ceram. Soc., 49,* 651 (1966).

CHAPTER 2

MODIFICATION OF SILICAGELS FOR CHROMATOGRAPHY

Walter A. Aue and Shubhender Kapila

Silicagels can be manufactured in a wide range of surface areas and pore volumes. They are quite inexpensive — despite some chromatographic evidence to the contrary — and serve in many applications as dehydrating agents, fillers, catalysts, and chromatographic phases.

For the latter use, the characteristics of silicagel surfaces are of paramount importance. While silicagels perform well in gas chromatographic separations of gases or nonpolar organics with low molecular weights, they would not be considered suitable for the analysis of more complex, polar compounds: the combined effects of high temperatures and a catalytic surface see to that. Silicagels are, of course, extensively used in liquid chromatography, but even under these mild conditions can the surface cause decomposition reactions to occur.

Because of the economic importance of silicagels, they rank among the more thoroughly investigated materials. Chromatographic retention can be predicted from surface areas and thermodynamic measurements; however, the overall chromatographic performance is all but inaccessible to theoretical treatment. Often, two batches of a particular silicagel — which shares this distinction with other adsorbents — will give rise to two distinct chromatograms, although other criteria may fail to reflect this difference.

Many effects combine to yield what we have chosen above to describe as "overall chromatographic performance," a term which is perhaps best reflected by the reaction of an experienced chromatographer upon inspection of the recorder tracing. It judges the sharpness, symmetry and relative position of peaks; takes note of how well trace amounts survive the tortuous path through the column, how far isomers are separated, and how strong a role solute polarity plays; and then weighs these characteristics in the context of instrumental settings, notably temperature and the nature and speed of the mobile phase.

This diversity of criteria is matched by the diversity of effects that influence the chromatographic performance of silicagels. They include, among others, the size, shape and distribution of pores, the chemical and physical nature of the surface, and the shape and size distribution of the particles. For instance, the presence of trace metals and the density of silanol groups on the surface, the relative speeds of solute migration inside *vs.* outside the particles, and the even less-understood factors involved in packing a column, all contribute to the picture. In addition, liquid chromatography obfuscates this picture by introducing a wide diversity of mobile phases, which can (conceptually at least) engage in different types of molecular interactions and mechanisms. All things considered, chromatographic separation still remains a wide-open field for the experimentalist.

Silicagel has been used for gas chromatography since approximately 1940,[1] succeeding, of course, its use in liquid chromatography. Since then, considerable efforts have been expended to improve the properties of this material for chromatography. Not only typical silicagels, but silicic supports as a whole have been the subject of renewed interest in recent years,[2-6] mainly owing to the fast-developing field of high-pressure liquid chromatography.

Considerable efforts have also gone into the characterization (and standardization) of silicagels in terms of their pore geometry and "activity."[7,8] It is possible to control, to some degree, the micro and macro structure of particles and the nature of their surfaces[5,6,9,10] by control of conditions during their synthesis.

This approach has its limitations, however, and quite a number of techniques have been developed to alter commercially available silicagels. Sometimes these techniques are classified as involving either geometric or chemical modifications of the surface. Although the distinction between the two is not always clear-cut, the former may include treatments with steam, water, or salts at high temperatures, while the latter would encompass such techniques as acid-washing, hydroxylation or dehydroxylation, chemical reactions involving silanol groups (esterification, silylation, Grignard reactions, etc.), and deposition of inorganic, highly polar, or polymeric substances on the surface. Presumably, geometric modifications change mainly surface area and pore size, whereas chemical modifications would alter the chemical nature and, consequently, the chromatographic "activity" of the surface available.

Treating silicagels with water and/or salts can bring about a drastic increase in pore size and, correspondingly, a decrease in surface area. Figure 2.1 illustrates this point with a comparison of silicagel Davidson Grade 62 before and after treatment with water at 380°. It seems likely, however, that such geometric modifications are in some cases accompanied by changes in the chemical nature of the surface — changes which may find their expression only in particular chromatographic separations.

Calcinations with various salts have been described in a rather broad-ranged German patent.[11] Owing mainly to the efforts of Kiselev and collaborators, the literature (the Eastern one, that is) appears replete with information on the effects of "hydrothermal treatment," which involves the use of superheated steam or, alternatively, water vapor at elevated temperatures and pressures.[12-21] Depending on temperature, pressure and time, the surface area of silicagels is drastically reduced by a widening of the pores, leading to much shorter chromatographic retention times. Total pore volume, however, changes little. The presence of salts in hydrothermal treatments can exert a strong accelerating influence on the reaction,[22,23] and similar, high-temperature treatments can be carried out with alcohols.[24] (Although the reaction with alcohols esterifies the silanol groups, these are easily recovered by hydrolysis.) Dry heat has also been employed, particularly to remove ultrafine pores by sintering.[25]

It is noteworthy that "geometrically modified" silicagels are rarely used as starting materials for further chemical modifications, such as the synthesis of bonded liquid phases. The use of commercially available silicic materials is convenient, to be sure, but it may not always represent the best choice of support.

In some work from our own laboratory, for example, heavy loads (up to 30%) of octadecylpolysiloxane were support-bonded to a commercial, wide-pore silicagel (Davidson Grade 62). The resulting materials still

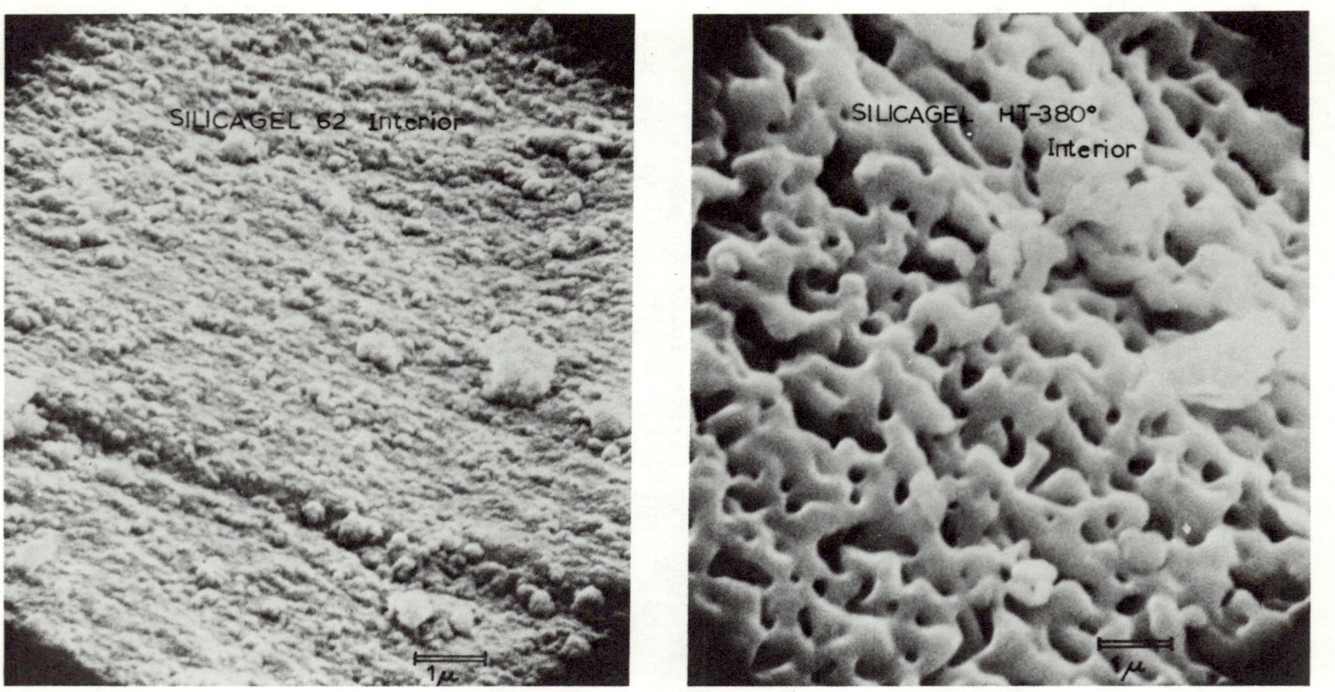

Figure 2.1. *Scanning electron micrographs of the interior of silicagel particles at 10,000 X magnification. Left: Silicagel Davidson Grade 62, acid-washed. Right: Same after overnight treatment with water at 380° (somewhat above the critical temperature).*

retained, in gas chromatography, some of the undesirably high surface activity of the original silicagel. Hydrothermal treatment (or more precisely, our version thereof, involving the liquid rather than the gas phase) reduced this activity considerably, although the silicagel surface still seemed to influence chromatographic performance.[26] The situation may be comparable, to some degree, to one described in a recent physicochemical study of an esterified surface[27] — and possibly to other bonded materials as well.

Chemical modification of silicic surfaces is, of course, much more familiar to chromatographers than the "geometric" variety discussed above. The simplest form of chemical modification is probably acid-washing — a procedure whose beneficial effects on succeeding chromatographies are attributed to the removal of iron and other impurities from the surface, and which most users of silicic supports, for better or for worse, leave to the manufacturer to execute.

The information which exists on acid-treatment of silicagels for gas chromatographic purposes suggests that this procedure is definitely advantageous. In the case of diatomaceous earth particles, of course, it is all but indispensable.[28-30]

Acid-washing is often followed by silylation, and quite a variety of chlorosilanes have been employed for this reaction. On diatomaceous earths, again, dimethyldichlorosilane is considered the best agent for deactivating the surface. Considering the small amounts of water necessary for formation of polysiloxanes, and the fact that a difunctional rather than a monofunctional compound turns out the best performance, this reaction may conceivably entail the formation of short silicone chains in some cases.

Support-bonded silicones represent, to some extent, an extension of this approach, although their syntheses are quite a bit more involved[31-34] and, in the case of recently marketed products, carefully shrouded by pretty chromatograms. The subject of support-bonded phases[35] is, of course, the main theme of this monograph and, being thus assured of in-depth coverage, need not be further discussed here. Suffice it to add to this review a few remarks, some of them speculative, to the subject of "chemically bonded" materials.

In early work on silicones in our laboratory, it soon became apparent that the main criterion for the "bonded" quality of phases was their behavior in an exhaustive extraction (which sometimes ran for weeks) in Goldfisch or Soxhlet equipment. Under these conditions, and with a variety of solvents, quite a number of silicones (not all of them of our own make) were removed quantitatively from the support. A short extraction, on the other hand, would have been entirely inadequate to remove the non-support-bonded material.

This fact is hardly surprising if one considers the capability of some silicones to rearrange, cross-link, and form close to insoluble rubbers. Although some information on the degree of cross-linking, and thus solubility, can presumably be obtained from an estimate of solute diffusion rates (using the C-term of the Van Deemter equation), it may be difficult at times to attribute the non-extractability of a particular polymer with certainty to true chemical bonding. It is conceivable in many cases that the polymer could simply be highly cross-linked and insoluble under the conditions employed. The peculiar chromatographic behavior of certain of the "bonded" silicones in liquid systems may, in fact, suggest just that.

This speculation does not apply, obviously, to some of the smaller molecules bonded in a "brush" type layer. In fact, the very ease with which some of them can be removed from the support by a solvent containing traces of water attests to their former true chemical bonding.

There are still other ways to modify and/or deactivate silicic surfaces — some of temporary, some of a more permanent nature. They share the approach to cover by polar molecules whatever "active sites" there are on the support. The molecules used in the cover-up vary considerably in size and influence, from precious metals,[36] to water in the carrier gas, to polymer chains deposited on the surface. They make little difference to *nonpolar* solutes, but are rather effective in reducing the energy of sorption, the nonlinearity of the sorption isotherm and, consequently, the peak asymmetry, of *polar* solutes. Extensive work has been done on these "tail-reducers" (see literature citations in references 12, 13, 29, 37 and 38).

In order for these compounds to be effective, they must, obviously, be selected with the support surface, the solute and, sometimes, a liquid phase in mind. Equally obvious is the fact that they need to be uniformly distributed on the support. The use of phosphoric acid for the gas chromatographic separation of volatile fatty acids may be cited here as one of many well-known examples.

The use of deactivating substances in liquid chromatography, however, presents a somewhat more difficult problem. Polar substances dissolved in the *mobile* phase shorten retention considerably by competing with the solute molecules, and polar substances deposited on the *stationary* phase often are of too high a solubility. One way of coping with the problem would seem to be the use of support-bonded polymers. What one may want, for instance, is a slightly modified, silica-type support with a vastly decreased tendency to adsorb, irreversibly, certain polar molecules, to decompose certain keto-steroids, or to denature certain proteins. A technique frequently employed in deactivating porous glasses for "gel" chromatography of biogenic substances consists of treating them with a solution of polyethyleneglycol. Apparently some of the polymer becomes quite strongly adsorbed (compare Reference 39).

The method is successful even though it is commonly accepted, without much question, that stationary liquid phases can be dissolved away from their supports. In fact, some gas chromatographic studies have relied on just such an extraction to measure the precise amount of liquid phase present on a particular packing. While the obtained values may be fairly correct for thick layers, thin films of a variety of polymers can behave in a quite different, and definitely surprising, manner.

This has been shown primarily for diatomaceous earths,[40–42] but silicagels seem to behave in a similar fashion, at least with Carbowax-20 M (a molecule formed from two polyethyleneglycol chains linked by a diepoxide, with a molecular weight between 15,000 and 20,000). Since some of our recent studies have been in this area, we shall make use of an author's prerogative and close this general review with a short discussion of our own work.

Our working hypothesis in these studies was that when a polymer chain of sufficient length is permitted to settle on a surface — in a manner which allows it to achieve a maximum of interaction with the surface and thus a minimum of potential energy — it will be difficult to remove by solvent action. This may, to some extent, be considered to parallel adhesion chemistry, where it is common practice to sum up all *inter*molecular forces between a surface and a molecule of adhesive, arriving at values in excess of those characteristic of *intra*molecular bonds.

The polymer molecules, in our case, are given the chance to arrange themselves on the surface in a heat treatment, not unlike the no-flow conditioning practiced in gas chromatography. Although a heat treatment seems to be the best way to achieve the desired effects,[42] less drastic approaches seem feasible also.[43] Our final step in producing physically bonded polymers is to subject the phases to several days of continuous extraction with boiling methanol or other appropriate solvents in order to remove the extractable bulk of the polymer. The "bonded" layer remaining has, on various occasions, survived several weeks of further extraction in Soxhlet equipment.

The evidence for physical — as opposed to chemical — bonding of the ultra-thin (1-20 Å) polymer films (on diatomaceous earths) is indirect: it is difficult to see how a diversity of liquid phases — including phases such as polyethylene, Apiezon L, dimethylpolysiloxane — could all react on or with the silicic surface and become insoluble. Chemical bonding of polyethyleneglycols, on the other hand, does appear possible under the chosen conditions (as in references 44 and 45), but the chemical bonds would not be expected to withstand days of exposure to methanol at boiling point temperatures.

Work on diatomaceous supports has further shown that the deactivation effected by a polymer film but a few Ångstroms thick is quite prominent: free cholesterol, for instance, can be gas-chromatographed in nanogram amounts. Retention times/temperatures are at a minimum and high carrier gas flows can be used. The materials are particularly well suited for isomer separations and can also serve as highly deactivated supports for other liquid phases. On silicagel surfaces, however, work has not yet gone beyond Carbowax 20 M layers and the chromatography of alcohols or phenols.

The gas chromatography of alcohols presents, of course, a severe test of a support's degree of deactivation. "Normal" silicagels do not allow small amounts of alcohols to pass through the GC column at all. However, a combination of deactivation methods — hydrothermal treatment (in the liquid phase), acid-washing, the deposition of nonextractable polymer, and humidification of the carrier gas[38] — will do the job quite effectively. The effects of these different methods are somewhat alike: Figure 2.2 shows a comparison based on the separation of n-alcohols on Silicagel 62 (hydrothermally treated at 260° and acid-washed) with water in the carrier gas at left

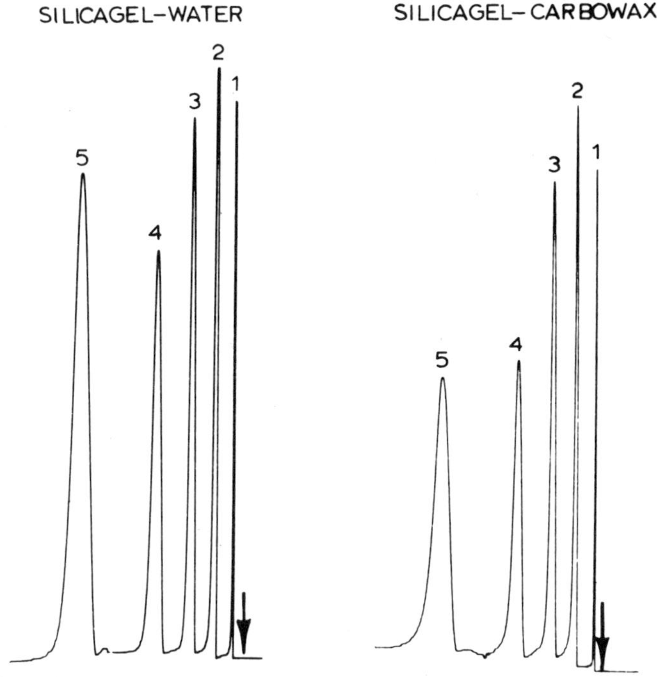

Figure 2.2. Gas chromatography at 140° of n-alcohols (methanol through pentanol) on Silicagel 62, hydrothermally treated at 280° overnight and acid-washed; packed in a 120 cm x 4 mm i.d. glass column. Left: Carrier gas (nitrogen) saturated, at room temperature, with water; bare silicagel. Right: Carrier gas dry; Silicagel coated with thin film (1-2 Å) of Carbowax 20 M.

and, at right, a one to two Ångstrom layer of Carbowax 20 M on the silicagel surface.

Surprisingly enough, it is possible to separate a number of alcohols even on bare, dry silicagel, as in the case of Figure 2.3 on a product obtained by hydrothermal treatment at 380°. This product is fully hydroxylated — a bit of information which derives its significance from the fact that, over the years, silanol groups have acquired a certain ill repute in gas chromatography and are commonly held responsible for a wide variety of deleterious effects. Here, they do not seem to detract much from the chromatographic performance.

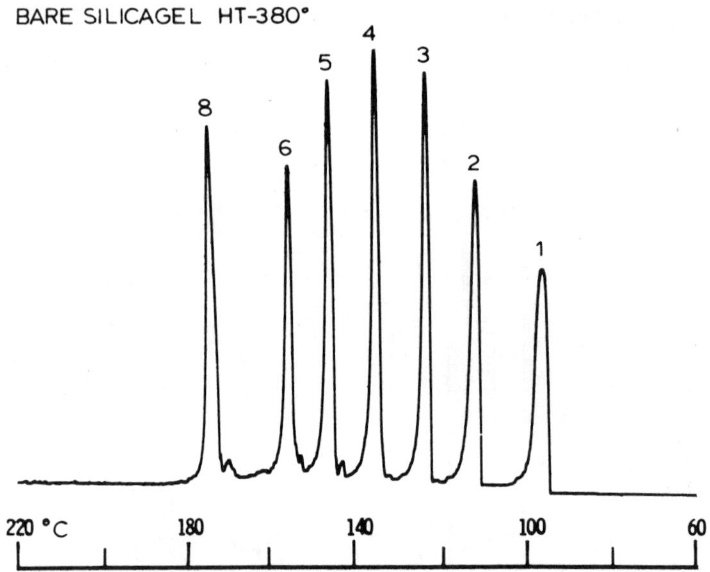

Figure 2.3. Temperature-programmed gas chromatography of n-alcohols (methanol to octanol) on bare silicagel, 40/60 mesh, derived from #62 by an overnight hydrothermal treatment at 380°, acid-washed, and packed in a 1.8 m x 2 mm i.d. glass column.

Figure 2.4 shows the further improvement brought about by the polymer layer: the chromatography now rivals the one obtainable on diatomaceous earths. This behavior is reflected in physicochemical data: the adsorption isotherms are almost linear over a wide range, and the heats of sorption for alcohols are not much different from those for hydrocarbons.[46]

Chromatographic effects in liquid chromatography are much more difficult to relate to defined parameters than those in gas chromatography, mainly owing to the much slower solute diffusion rate and its various consequences. Our rather crude studies in this area made use of the inexpensive, wide-pore silicagel Davidson Grade 62, with a surface area of 340 m^2/g and an average pore diameter of 170 Å.[47] This pore size would generally be regarded as large for liquid-solid chromatography. Yet, the 140-200 mesh fraction of Davidson Grade 62 improved markedly in chromatographic efficiency (a drop in HETP from 3.3 to 1.3 mm) when subjected to hydrothermal treatment at 260°.[41]

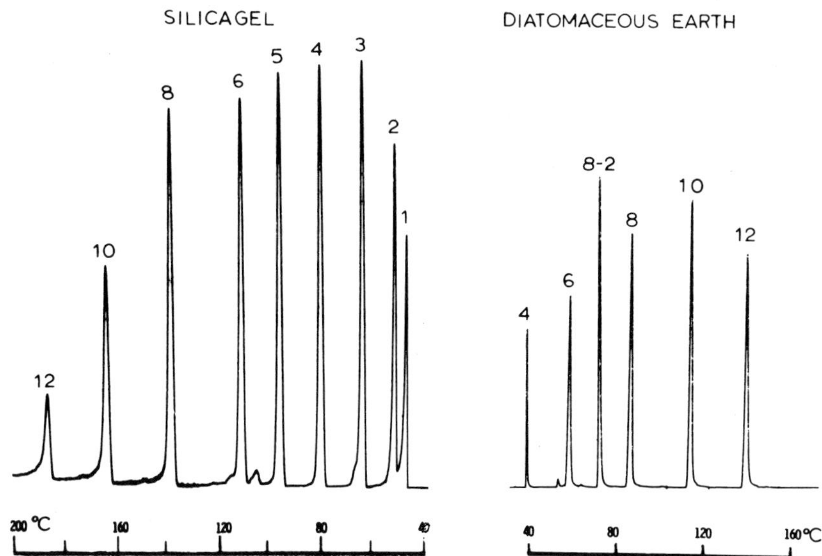

Figure 2.4. Temperature-programmed gas chromatography of n-alcohols (including 2-octanol) as defined by their carbon number. Stationary phases: Left: Approx. 6 Å layer of Carbowax 20 M on Silicagel 62, 40/60 mesh, hydrothermally treated overnight at 340° and acid-washed, packed in a 1.8 m x 2 mm i.d. glass column. Right: <15 Å layer of Carbowax 20 M on acid-washed Chromosorb W, 100/120 mesh, packed in a 5 ft x 4 mm i.d. glass column.

(This treatment could be expected to approximately halve the available surface area and double the average pore diameter.) The deposition of nonextractable Carbowax 20 M brought about a further slight improvement.

It is not completely understood at present whether these effects in LC are due to changes in surface activity, or ease of packing, or improved mass transfer, or increased mobility of the solvent through the particles — or perhaps a combination of these (and other) effects. Such conceptual uncertainties can, of course, be clarified, but only at the cost of considerable experimentation. Furthermore, it is not at all clear how far the implications of these experiments extend in regard to other types of silicagels, polymers, or chromatographic systems.

For this review, may it suffice to assume that combined methods of geometric and chemical modifications of silicagels can bring about significant improvements in both gas and liquid chromatographies. Furthermore, these modifications appear to present a rather rewarding challenge to the experimentalist.

As in the oral presentation of this review may we be permitted to add, tongue partially in cheek, three useful bits of "chemthink" which have evolved from our work with silicagels and which, we admit, are somewhat at variance with current chromatographic concepts. First, silanol groups are not all that bad. Second, physical support-bonding can be stronger than chemical support-bonding. And third (judging from the looks of chromatograms), there is no difference anymore between gas-solid and gas-liquid chromatography.

ACKNOWLEDGMENT

Most of our own research as mentioned in this review was supported by EPA Grant R801050 and a grant from the Rockefeller Foundation. We express our sincere appreciation to Dr. Corazon R. Hastings, whose studies of polymer layers on diatomaceous earth surfaces have preceded similar approaches to the modification of silicagels, and who has been of invaluable help to the silicagel project.

REFERENCES

1. Ettre, L. S. *American Laboratory,* 10 (October 1972).
2. Majors, R. E. *American Laboratory,* 27 (May 1972).
3. Leitch, R. E. and J. J. De Stefano. *J. Chrom. Sci., 11,* 105 (1973).
4. Halpaap, H. *J. Chrom., 78,* 77 (1973).
5. Unger, K., J. Schick-Kalb and K.-F. Krebs. *J. Chromatog., 83,* 5 (1973).
6. Iler, R. K., H. J. McQueston and J. J. Kirkland. (As cited in J. J. Kirkland) *J. Chromatog., 83,* 149 (1973).
7. Halpaap, H. *J. Chromatog., 78,* 63 (1973).
8. Snyder, L. R. *Separation Science, 1,* 191 (1966).
9. Zhdanov, S. P., A. V. Kiselev, E. V. Koromal'di, G. L. Kustova, B. A. Lipkind, and Yu. S. Nikitin. in *Adsorbenty,* 3rd ed. 1969, 44-7, Nauka, Leningrad; as cited in *C.A., 77,* 154546J.
10. Girgis, B. S. *J. Appl. Chem. Biotechn., 22,* 905 (1972).
11. Krebs, K. F. *Ger. Offen.* 2,042,910 (March 1972); *C.A. 77*, 22374S.
12. Kiselev, A. V. and Y. I. Yashin. *Gas-Adsorption Chromatography*, (New York: Plenum Press, 1969).

13. Kiselev, A. V. *J. Chromatog.*, *49*, 84 (1970)
14. Kiselev, A. V. and Yu. S. Nikitin. *Sovrem. Probl. Fiz Khim.*, *3*, 195 (1968); *C.A.*, *71*, 64436.
15. Kiselev, A. V., Yu. S. Nikitin and E. B. Oganesyan. *Colloid J.*, *31*, 416 (1969).
16. Basiljeva, V. C., I. V. Drogaleva, A. V. Kiselev, A. Ya. Korolev and K. D. Sherbakova. *Dokl. Akad. Nauk. SSSR*, *136*, 852 (1961).
17. Akshinskaya, N. V., V. Ya. Davydov, A. V. Kiselev and Yu. S. Nikitin. *Colloid J.*, *28*, 1 (1966).
18. Akshinskaya, N. V., B. E. Beznogova, A. V. Kiselev and Yu. S. Nikitin. *Zh. Fiz. Khim*, *36*, 2277 (1962).
19. Kiselev, A. V., Yu. S. Nikitin, A. I. Sarakhove and E. B. Oganesyan. *Kolloid Zh.*, *30*, 842 (1968); *C.A.*, *70*, 4155k.
20. Bebris, N. K., A. V. Kiselev and Yu. S. Nikitin. *Kolloid Zh.*, *29*, 326 (1967).
21. Janak, J. and R. Staszewski. *J. Gas. Chrom.*, 47 (February 1964).
22. Hill, T. *German Offen.* 1,958,312 (1970); *C.A.*, *74*, 45977.
23. Kiselev, A. V., Yu. S. Nikitin and E. B. Oganesyan. *Colloid J.*, *28*, 537 (1966).
24. Waksmundzki, A., R. Leboda and Z. Suprynowicz. *Chemia Analityczna*, *18*, 727 (1973).
25. Akshinskaya, N. V., T. A. Baigubekova, A. V. Kiselev and Yu. S. Nikitin. *Colloid J.*, *28*, 139 (1966).
26. Tsai, P. Ph.D. Thesis, University of Missouri, Columbia (1972).
27. Uihlein, M. and I. Halász. *J. Chrom.*, *80*, 1 (1973).
28. Ottenstein, D. M. *J. Gas Chrom.*, *1*, 11 (1963).
29. Ottenstein, D. M. *J. Chrom. Sci.*, *11*, 136 (1973).
30. Bowman, M. C. and M. Beroza. *J. AOAC*, *54*, 1086 (1971).
31. Abel, E. W., F. H. Pollard, P. C. Uden and G. Nickless. *J. Chrom.*, *22*, 23 (1966).
32. Stewart, H. N. M. and S. G. Perry. *J. Chrom.*, *37*, 97 (1968).
33. Bossart, C. T. *ISA Trans.*, *7*, 283 (1968), U.S. Patent 3,514,925 (1970).
34. Aue, W. A. and C. R. Hastings. Midwest Regional ACS Meeting, Manhattan, Kansas, October 1968. U.S. Patent 3,666,530 (1972).
35. Locke, D. C. *J. Chrom. Sci.*, *11*, 120 (1973).
36. Ormerod, E. C. and R. P. W. Scott. *J. Chrom.*, *2*, 65 (1959).
37. Kiselev, A. V., N. V. Kovaleva and Yu. S. Nikitin. *J. Chrom.*, *58*, 19 (1971).
38. Aue, W. A., S. Kapila and K. O. Gerhardt. *J. Chrom.*, *78*, 228 (1973).
39. Iler, R. K. *The Colloid Chemistry of Silica and Silicates.* (Ithaca, N.Y.: Cornell University Press, 1955), p. 58.
40. Aue, W. A., C. R. Hastings and S. Kapila. *J. Chrom.*, *77*, 199 (1973).

41. Aue, W. A., C. R. Hastings and S. Kapila. *Anal. Chem.*, *45*, 725 (1973).
42. Hastings, C. R., J. M. Augl, S. Kapila and W. A. Aue. *J. Chromatog.*, *87*, 49 (1973).
43. McCullough, P. Master's Thesis, University of Missouri, Columbia (1974).
44. Iler, R. K. U.S. Patent 2,657,149 (1952).
45. Halasz, I. and I. Sebestian. *Angew. Chemie (Intl. Ed.)*, *8*, 453 (1969).
46. Kapila, S., W. A. Aue and J. M. Augl. *J. Chrom.*, *87*, 35 (1973).
47. Klein, P. D. *Anal. Chem.*, *34*, 733 (1962).

CHAPTER 3

CHEMICALLY BONDED STATIONARY PHASES FOR REVERSED-PHASE PARTITION CHROMATOGRAPHY

Reed C. Williams

INTRODUCTION

The ease and utility of high speed liquid chromatography has been greatly advanced by the introduction of chemically bonded stationary phases.[1,2] Before the introduction of these bonded packings, the stationary phase of partition and ion exchange column packings had to be either a liquid or a polymer coated on the surface of the pellicular packing support. The chemically bonded packings are far more durable and stable with column life lasting six months to a year while the coated packings often last only a few weeks.

The chemically bonded packings are less efficient than the corresponding liquid coated columns. However, the chemically bonded packings are much more efficient than corresponding polymer coated packings[3] and this can be seen in Figure 3.1. Two chromatograms are shown. Both show the separation of some fused ring aromatics with the same mobile phase on two similar column packings. The top chromatogram shows the separation of the mixture on a 2.1 mm by one meter column of Zipax® HCP in a mobile phase of 40% H_2O and 60% methanol at a temperature of 40°C. The Zipax HCP consists of a hydrocarbon polymer which is coated onto the surface of the Zipax column support. The bottom chromatogram shows the separation of a similar mixture on a column of Permaphase® ODS with the same mobile phase and at 50°C. Permaphase ODS contains an octadecyl hydrocarbon group chemically bonded onto the surface of the Zipax support and is

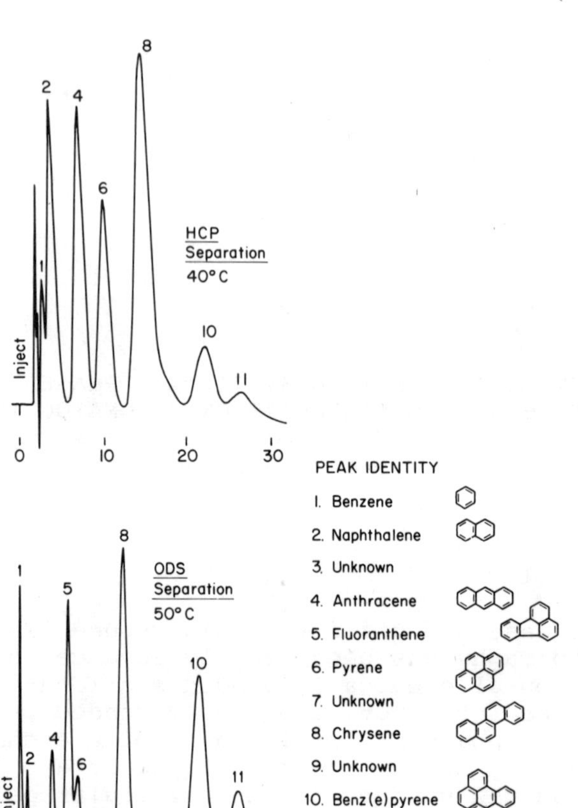

Figure 3.1. Comparison of the efficiency of Zipax HCP and Permaphase ODS.
Column: one meter x 2.1 mm I.D.
Mobile phase:-60% water/40% methanol (V/V)
Flow: 1 ml/min
Detector: UV photometer at 254 nm.

similar in nature to the hydrocarbon polymer. The greater efficiency of the chemically bonded support is obvious. This is apparently caused by the fact that the stationary phase can be more evenly bonded across the surface of the packing than coated, and this uniform layer gives greater efficiency. Similar efficiency increases were also found for chemically bonded ion exchange columns when compared to the resin coated pellicular packings.[4]

Because of their increased thermal stability these chemically bonded packings could also be used at elevated temperatures without column degradation. This was particularly valuable for those separations which used the more viscous higher boiling solvents, such as water, for the mobile phase. As the column temperature increased, the mobile phase viscosity decreased and this resulted in faster mass transfer and increased column efficiency. This efficiency increase with temperature increase has been observed for both reversed phase partition[3] and ion exchange with chemically bonded pellicular packings.[4]

The introduction of the chemically bonded packings made gradient elution chromatography a practical reality for both partition and ion exchange modes of liquid chromatography. With the liquid and polymer coated packings, one dared not to change the mobile phase very much for fear of stripping the stationary phase from the packing. Since this fear was unnecessary with the bonded packings, gradient elution of the mobile phase became an accepted technique.

An example of the power and usefulness of the gradient elution technique is shown in Figure 3.2. A sample containing several different components was chromatographed. The HMX, RDX, DNT and TNT are all nitrated hydrocarbons which are quite water soluble and are barely retained on a column of Permaphase ODS in a water mobile phase. The dibutylphthalate (DBP) and diphenylamine (DPA) are much less water soluble and would be strongly retained on the same column in a water mobile phase although unretained in a methanol mobile phase. In order to analyze all of these components in a single run, it is necessary to program the mobile phase from water to methanol. This allows all components to be retained for separation but also maintains a reasonable analysis time.

REVERSE-PHASE PARTITION

The octadecyl or C_{18} bonded packing used in reverse-phase partition has been the most useful of the chemically bonded column packings now used in liquid chromatography. With this reversed-phase technique the mechanism of separation is the partition of a relatively nonpolar sample between the

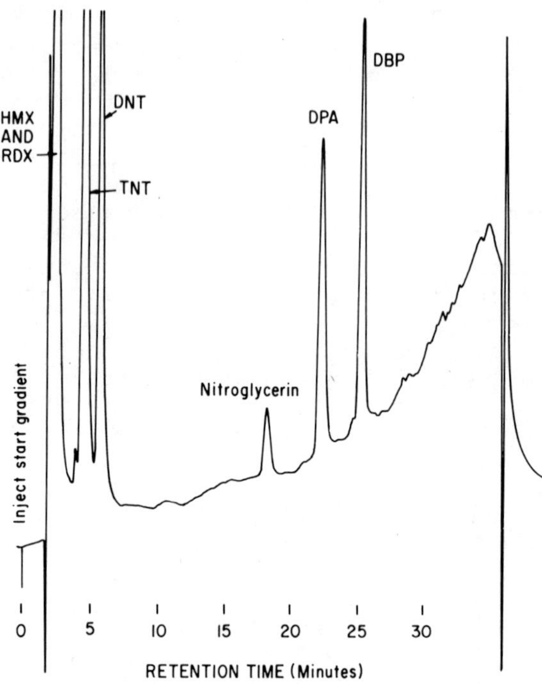

Figure 3.2. Separation of explosives on Permaphase ODS. Column: Permaphase ODS (one meter x 2.1 mm I.D.) Mobile phase: exponential gradient from water to methanol at 3%/min Flow rate: 1 ml/min Detector: UV photometer at 254 nm.

polar mobile phase and the nonpolar stationary phase. This technique was first introduced by Howard and Martin in 1950,[5] but was not widely used in liquid chromatography because of the instability of the column packings. However, with the introduction of the octadecyl bonded column packing in 1970, this reversed-phase method became widely used. It proved to be an efficient and reproducible separation technique which was readily adaptable to gradient elution of the mobile phase.

The reversed-phase method has been successful and several papers using this technique have been published. However, not all likely samples are adequately separated on the standard octadecyl column packing. An example is shown in Figure 3.3. The chromatogram on the right shows the incomplete separation of diaminoterephthalic acid from impurities on a column of Permaphase ODS. The mobile phase was 25% THF and 75% water with a small amount of phosphoric acid. The separation of the main component and impurities is not adequate and was not improved as the retention was increased. The chromatogram on the left shows a superior separation on an aromatic bonded column in the same mobile phase. This column packing contains phenyl groups bonded to the support.

This was dramatic evidence of the fact that other chemically bonded column packings, besides the octadecyl, can be used with good effort in reversed-phase partition. Since the reversed-phase method is such a versatile technique, it was decided to pursue

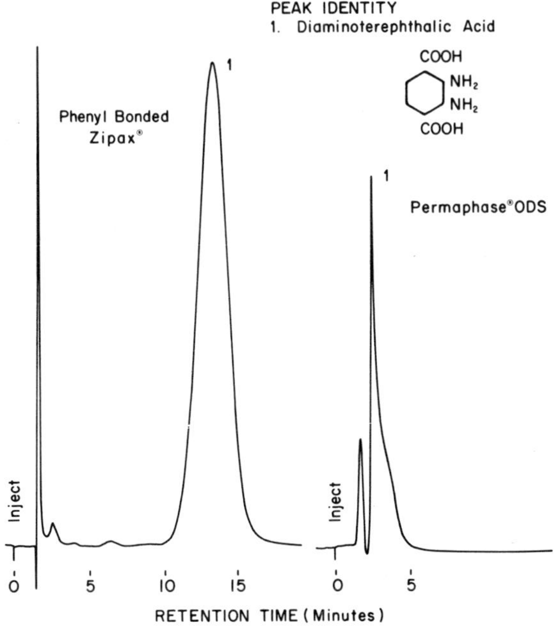

Figure 3.3. *Separation of diaminoterephthalic acid on Permaphase ODS and phenyl bonded Zipax.*
Columns: one meter x 2.1 mm I.D.
Mobile phase: 75% H_2O, 25% THF, 0.01% H_3PO_4
Temperature: 24°C
Flow: 1 ml/min
Detector: UV photometer

a study of the selectivity differences of bonded column packings which might be useful for reversed-phase partition. To do this, a test mixture of naphthalene derivatives of different functional groups was chosen. These functional groups included an amine, nitrile, methoxy, methyl and chloro. The mixture was used to test column packings for selectivity differences. A number of practical mobile phases such as water-methanol or water-acetonitrile could have been used for this study. However, our experience has been that different mobile phases will not greatly change the column selectivity for different functional groups in reverse phase partition. Because we were most familiar with water methanol, we chose this as a mobile phase.

An example of the separation of the test mixture is shown in Figure 3.4. The top chromatogram shows

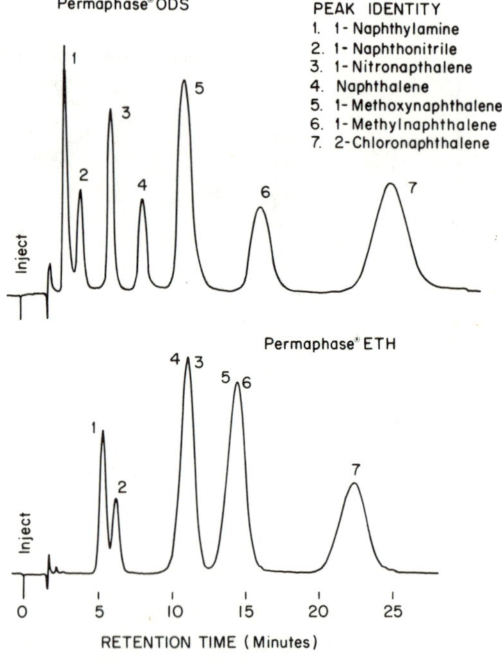

Figure 3.4. Separation of naphthalene derivatives on Permaphase ODS and ETH.
Columns: Permaphase ODS and ETH (one meter x 2.1 mm I.D.)
Mobile Phase: 40% Methanol/60% H_2O
Temperature: 24°C
Pressure: 1500 psi
Flow: 1.0 ml/min
Detector: UV photometer.

the test mixture separated on a column of Permaphase ODS and the bottom chromatogram on Permaphase ETH. The ETH column contains aliphatic ether groups bonded to the Zipax.[2] The mobile phase was 40% methanol and 60% water for both separations. The retention times of the compounds are similar for both columns. However, there are some interesting differences in selectivity between the two. The components are all clearly separated on the ODS but the selectivity of the ETH is such that some of the components are not resolved. In particular, nitronaphthalene (component 3) and methoxynaphthalene (component 5) are more strongly retained on the ETH than the other components.

Another comparison of selectivity differences between the two column packings is shown in Figure 3.5 where a series of benzene derivatives was chromatographed in a water mobile phase. Generally, the derivatives are more strongly retained on Permaphase ETH, but this is most obvious for phenol, methyl benzoate and nitrobenzene. Although these

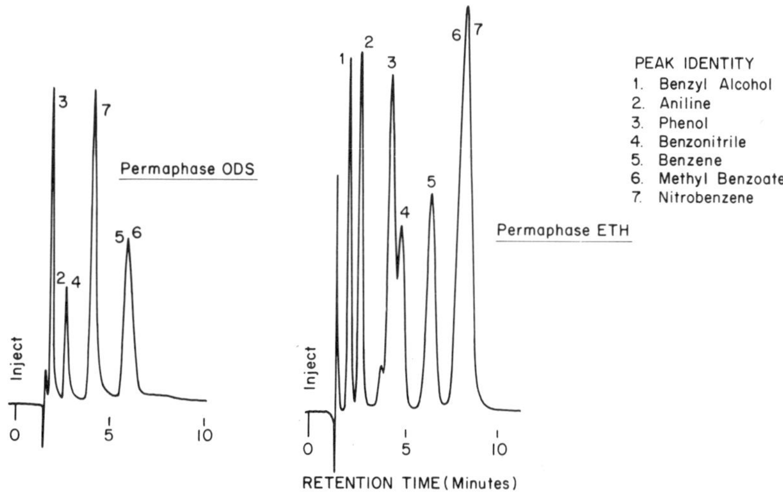

Figure 3.5. Separation of benzene derivatives on Permaphase ETH and ODS
Columns: Permaphase ODS and ETH (one meter x 2.1 mm I.D.)
Mobile phase: water
Temperature: 24°C
Pressure: 1000 psi
Flow: 1.2 ml/min
Detector: UV photometer.

selectivity differences may not be large, they can be of importance when analyzing some mixtures as will be shown later in a practical example.

The study continued on to compare differences between the ODS and the experimental aromatic bonded Zipax as shown in Figure 3.6. Again, the test mixture was chromatographed on both columns under similar conditions. The phenyl column turned out to be less retentive so the mobile phase was changed

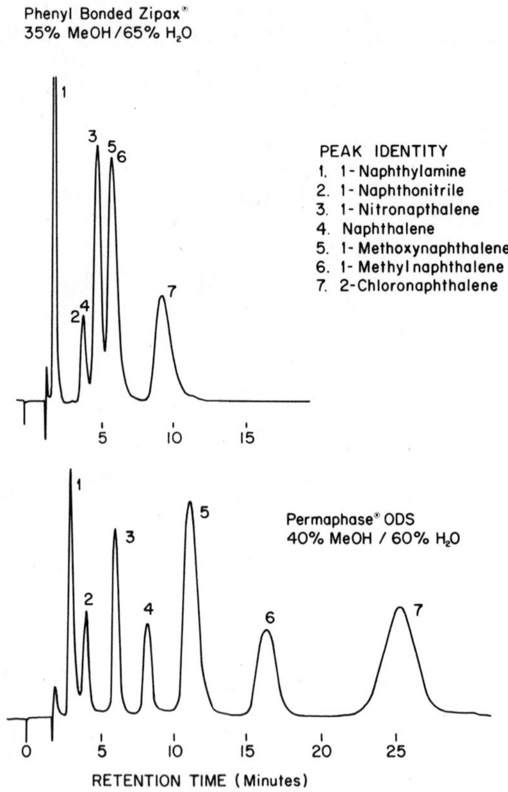

Figure 3.6. Separation of naphthalene derivatives on Permaphase ODS and phenyl bonded Zipax.
Columns: one meter x 2.1 mm I.D.
Temperature: 24°C
Pressure: 1500 psi
Flow: 1.0 ml/min
Detector: UV photometer.

slightly in order to give comparable retention. The relative elution order of nitronaphthalene and naphthonitrile was greater on the phenyl bonded Zipax than the ODS in comparison to the other components.

Figure 3.7 shows chromatograms of the test mixture on columns of ODS and on an experimental

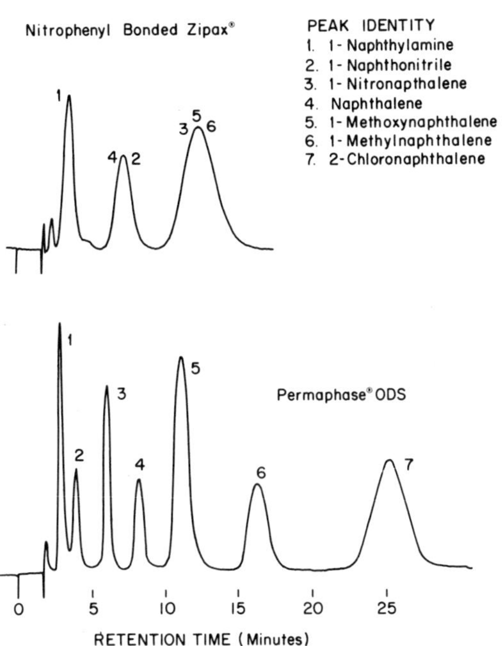

Figure 3.7. Separation of naphthalene derivatives on Permaphase ODS and nitrophenyl bonded Zipax.
Columns: one meter x 2.1 mm I.D.
Mobile phase: 40% methanol and 60% H O
Temperature: 24°C
Pressure: 1500 psi
Flow: 1.0 ml/min
Detector: UV photometer.

nitrophenyl bonded Zipax with a mobile phase of 40% methanol and 60% water. The nitrophenyl column was not very efficient, but again there were significant selectivity differences. Both the nitronaphthalene and naphthonitrile were more strongly retained on the nitrobenzyl in comparison to the other components.

This comparison study of column packing was interesting and the information on column selectivity differences was put to practical use on an applications problem. We were interested in separation and analysis of some nitrated aromatic molecules used in explosive propellant mixtures. The requirements of the sample preparation and analysis were such that a reversed phase separation was desirable. Figure 3.2 shows an example of this separation done on a column of Permaphase ODS. The mobile phase was changed from water to methanol by gradient elution. The nitrated compounds, TNT, DNT, RDX and HMX, were not strongly retained on the ODS column, and this caused serious problems with their analysis. Even much longer columns of ODS did not solve the problem.

From the selectivity experiments we knew that nitrated materials were more strongly retained on Permaphase ETH. So in order to solve our separations, we then took both a 1/2 meter column of Permaphase ETH and a 1-1/2 meter column of Permaphase ODS and ran the same propellant mixture with the same mobile phase. The results are shown in Figure 3.8. The samples can be completely separated with this column combination so that quantitative analysis of the various components can be obtained. This is an example of how column selectivity differences in reversed-phase chromatography can be put to use to solve practical applications problems.

In conclusion, because of their efficiency, durability and application to a wide variety of materials, the octadecyl bonded packings will be the column of first choice for most reversed-phase applications. However, it is obvious that other column packings besides the octadecyl bonded packings may be used effectively in reversed-phase partition chromatography. Hopefully, in the future a number of chemically bonded stationary phases will be available for reversed-phase partition which will allow the chromatographer greater flexibility in altering column selectivity for resolution of difficult separations problems.

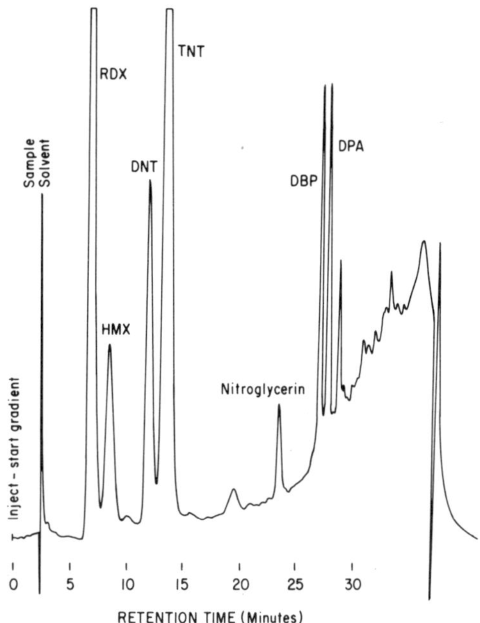

Figure 3.8. *Separation of explosives using columns coupled in series.*
Columns: Permaphase ODS (1-1/2 meter x 2.1 mm I.D.) and Permaphase ETH (1/2 meter x 2.1 mm I.D.)
Mobile phase: exponential gradient from water to methanol at 3%/min
Flow: 1 ml/min
Detector: UV photometer

REFERENCES

1. Halász, I and I. Sebastian. *Angew Chem. Interat. Ed., 8,* 453 (1969).
2. Kirkland, J. J. and J. J. DeStefano. *J. Chromatog. Sci., 8,* 309 (1970).
3. Schmit, J. A., R. A. Henry, R. C. Williams, and J. F. Dieckman. *J. Chromatog. Sci., 9,* 645 (1971).
4. Henry, R. A., J. A. Schmit, and R. C. Williams. *J. Chromatog. Sci., 11,* 358 (1973).
5. Howard, G. A. and A. J. P. Martin. *Biochem. J., 56,* 539 (1950).

CHAPTER 4

PERMANENTLY BONDED STATIONARY PHASES IN GAS AND LIQUID CHROMATOGRAPHY

James N. Little, Dennis F. Horgan, Jr., Richard L. Cotter, and Richard V. Vivilecchia

This chapter reviews work initiated in 1968 concerning permanently bonded stationary phases in both gas and liquid chromatography.

1. the effect of the length of the bonded phase
2. the efficiency of bonded phases versus conventionally coated phases
3. a comparison of aliphatic and aromatic phases in reverse phase liquid chromatography.

In addition, the review discusses numerous applications using bonded phases and how one chooses the proper bonded phase for a particular problem. It is divided into two sections, gas chromatography and liquid chromatography packings.

GAS CHROMATOGRAPHY

Our work involving bonded gas chromatography packings involves reacting alcohols to silica surfaces as shown in Figure 4.1 (commercial name, DURAPAK*). To develop a series of different polarity packings, three alcohols were bonded onto silica:

---

*Registered trademark of Waters Associates, Inc.

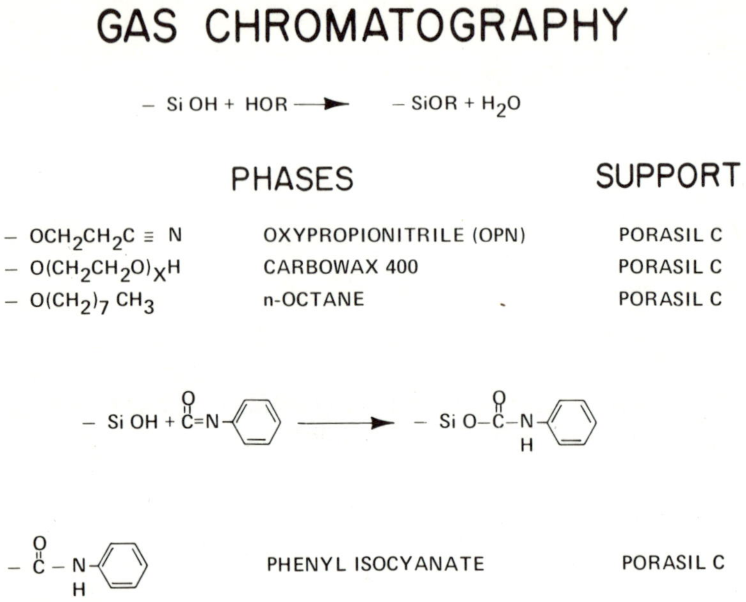

GAS CHROMATOGRAPHY

$$-SiOH + HOR \longrightarrow -SiOR + H_2O$$

	PHASES	SUPPORT
$-OCH_2CH_2C \equiv N$	OXYPROPIONITRILE (OPN)	PORASIL C
$-O(CH_2CH_2O)_XH$	CARBOWAX 400	PORASIL C
$-O(CH_2)_7CH_3$	n-OCTANE	PORASIL C

$$-SiOH + O{=}C{=}N{-}C_6H_5 \longrightarrow -SiO{-}C(=O){-}NH{-}C_6H_5$$

$-C(=O)-NH-C_6H_5$	PHENYL ISOCYANATE	PORASIL C

Figure 4.1. Summary of gas chromatography packings.

1. Hydroxypropionitrile (OPN)
2. Carbowax 400 (polyethylene glycol of approximately 400 MW)
3. Octanol.

A special purpose packing was also prepared by reacting phenyl isocyanate onto the silica surface (Figure 4.1). This bond is a urethane bond as opposed to the ester bonds described above. For all of these packings, the support chosen was PORASIL C. PORASIL C is a spherical, porous silica packing of 50-100 m^2/g surface area. After the reaction onto PORASIL, the amount of bonded phase is approximately 3-5%. Higher and lower surface area supports were tested which resulted in higher and lower amounts of bonded phase, respectively. PORASIL C appeared to be the right compromise in efficiency, loadability and retention.

The optimum velocity for the DURAPAKs has been found to be between 6-10 cm/sec (or approximately 60-100 ml/min for a 1/4 inch OD column). The packings all have a long flat minimum region in the

H/u *vs.* u curve (Figure 4.2) and can be operated at flow rates substantially higher than optimum with little loss in efficiency.[1]

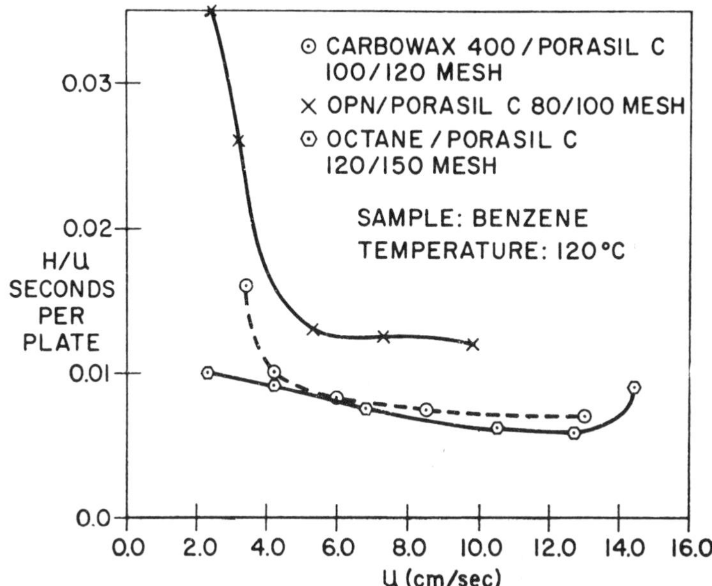

Figure 4.2. H/U vs. u. (Reprinted with permission from J. Chromatographic Science.)

Of the three packings, the relative retention of polar solutes is highest on the n-octane packings. This is a result of some residual Si-OH groups remaining on the surface of this packing and adsorption effects contributing to the separation. For hydrocarbon solutes, the relative retention is approximately the same on all three packings. For aromatic compounds, the relative retentions on OPN and n-octane are similar. In relating the elution behavior of representative compounds, the Carbowax 400-DURAPAK is the least polar packing, the OPN-DURAPAK is of intermediate polarity, and the n-octane-DURAPAK is the most polar.

Figure 4.3 dramatically shows the differences in polarity between the Carbowax 400 and OPN-DURAPAK packings. Acetone, which elutes very early on Carbowax 400, elutes after chloroform, benzene and

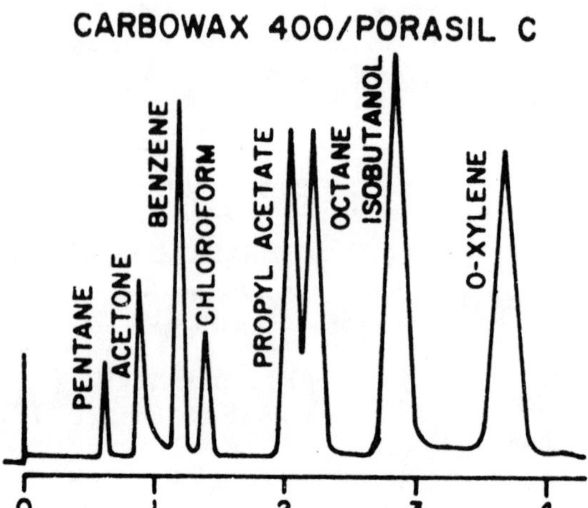

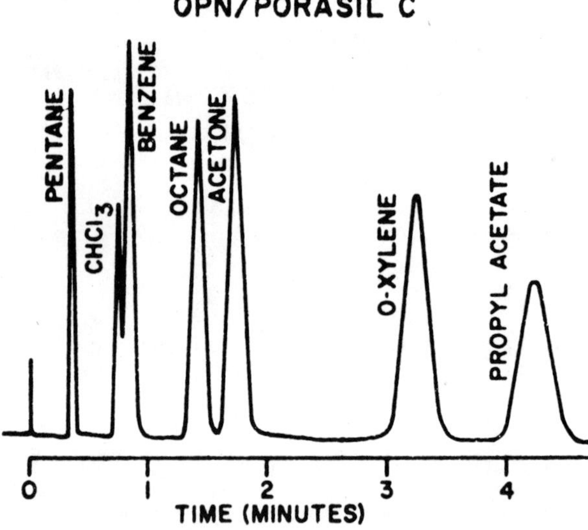

Figure 4.3. Polarity of DURAPAK columns. (Reprinted with permission from J. Chromatographic Science.)

octane on OPN. Propyl acetate, which elutes before octane on Carbowax 400-DURAPAK, elutes well after octane on the OPN-DURAPAK.

Figure 4.4 shows retention data, relative to n-octane, of a few compounds and confirms the order of polarity as CW400, OPN and n-octane.

POLARITY OF DURAPAK COLUMNS

150° C

COMPOUND	RELATIVE RETENTION		
	CARBOWAX 400/ PORASIL C	OPN/PORASIL C	OCTANE/ PORASIL C
FORMALDEHYDE	0.388	0.536	1.03
BUTYRALDEHYDE	0.544	1.52	1.58
ACETONE	0.449	1.21	2.23
2 – BUTANONE	0.612	1.79	3.37
METHYL ACETATE	0.440	0.912	1.75
ISOPROPYL ACETATE	0.768	2.17	5.00

Figure 4.4 Selective relative retention data on DURAPAK columns (relative to n-octane).

The important separation of C_4 hydrocarbon isomers has been obtained on a DURAPAK-n-octane/PORASIL C column at room temperature in only four minutes (Figure 4.5). The special purpose DURAPAK-Phenyl Isocyanate/PORASIL C was found to separate the C_1-C_3 hydrocarbon isomers in 2.5 minutes at room temperature (Figure 4.6).

Different molecular weight Carbowaxes were bonded to PORASIL to determine the effect of "brush" length. Relative retention data for three compounds (acetone, benzene and propyl acetate) are tabulated in Figure 4.7. All three compounds gave lower relative retention values on the low molecular weight Carbowaxes (200, 400 and 1000) than on the high molecular weight Carbowaxes (4000 and 20,000). Thus, the low molecular weight bonded Carbowaxes are relatively nonpolar when compared to the high molecular weight bonded Carbowaxes. The opposite polarity effect is found when these Carbowaxes are conventionally coated onto PORASIL.

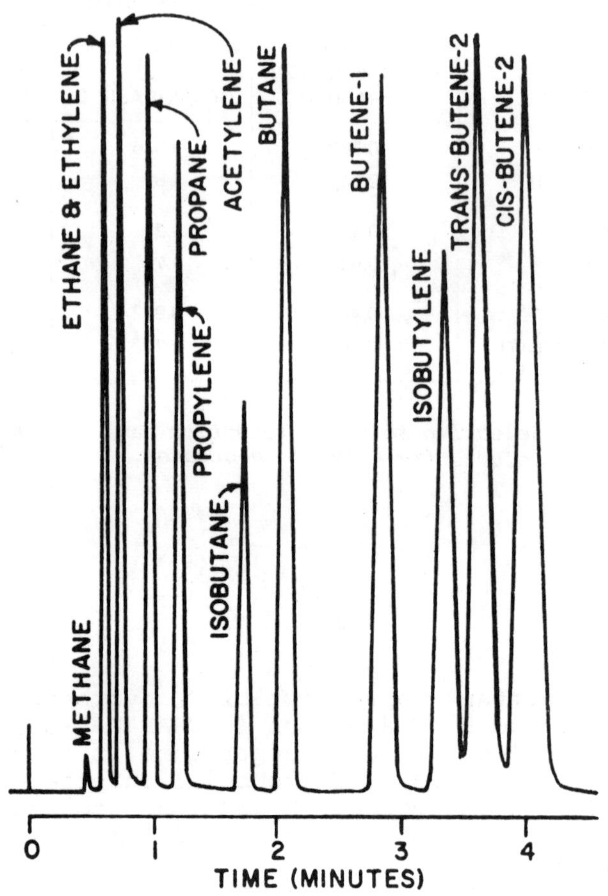

Figure 4.5. Separation of C_4 isomers on n-octane/Porasil C
Column: 1.5 meter x 2.3 mm ID
Temperature: 25°C
Flow Rate: 20 ml/min (N_2)
(Reprinted with permission from J. Chromatographic Science.)

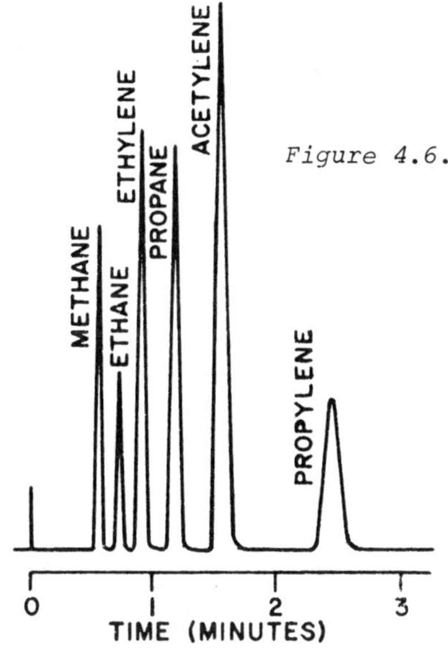

Figure 4.6. *Separation of C_1-C_3 isomers on phenyl isocyanate/Porasil C.*
Column: 1 meter x 2.3 mm ID
Temperature: 23°C
Flow Rate: 10 ml/min (N_2)
(Reprinted with permission from J. Chromatographic Science.)

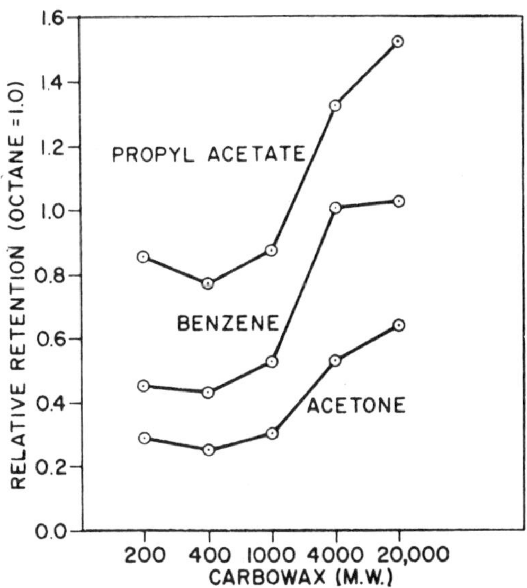

Figure 4.7. *Effect of Carbowax chain length.*
All columns: 1 meter x 2.3 mm ID
Temperature: 120°C
Flow Rate: 25 ml/min (N_2)
(Reprinted with permission from J. Chromatographic Science.)

LIQUID CHROMATOGRAPHY

Our work in liquid chromatography paralleled that in gas chromatography and involved bonding alcohols to silica surfaces as shown in Figure 4.8. The PORASIL C used was in the particle size range of 37-75 microns which was more optimum for high-pressure liquid chromatography. Besides PORASIL C, Carbowax 400 was also bonded onto a highly efficient, pellicular packing, CORASIL.[2]

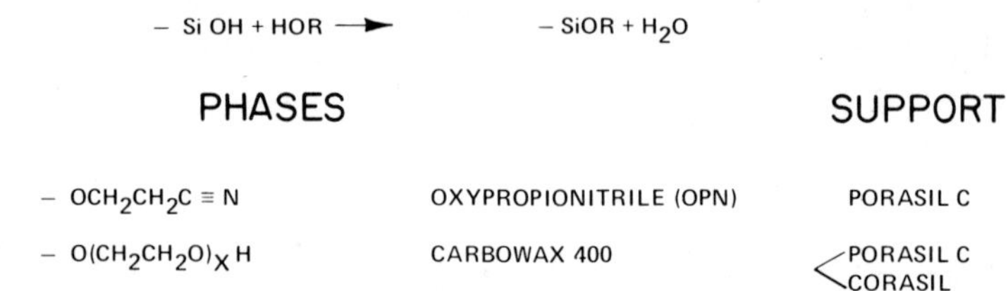

Figure 4.8. Summary of liquid chromatography packings (ester bonds).

As was done in gas chromatography, a comparison of chemically bonded and conventionally coated supports were evaluated.[3] Figure 4.9 lists the supports which were compared. In each case, the

PACKINGS EVALUATED

CONVENTIONALLY - COATED	*CHEMICALLY - BONDED*
1 % CARBOWAX 400/CORASIL	CARBOWAX 400/CORASIL
1 % CARBOWAX 4000/CORASIL	CARBOWAX 4000/CORASIL
1 % DIETHYLENE GLYCOL/CORASIL	DIETHYLENE GLYCOL/CORASIL
5 % CARBOWAX 400/PORASIL C	CARBOWAX 400/PORASIL C
33 % CARBOWAX 400/PORASIL C	

Figure 4.9. Types of packings evaluated. (Reprinted with permission of J. Chromatographic Science.)

weight per cent of stationary phase was the same for the chemically bonded and the conventionally coated supports

A four-component mixture consisting of (a) benzene, (b) nitrobenzene, (c) decanol, and (d) benzyl alcohol was used for evaluation of all packings. A separation of this mixture on DURAPAK-Carbowax 400/CORASIL is shown in Figure 4.10.

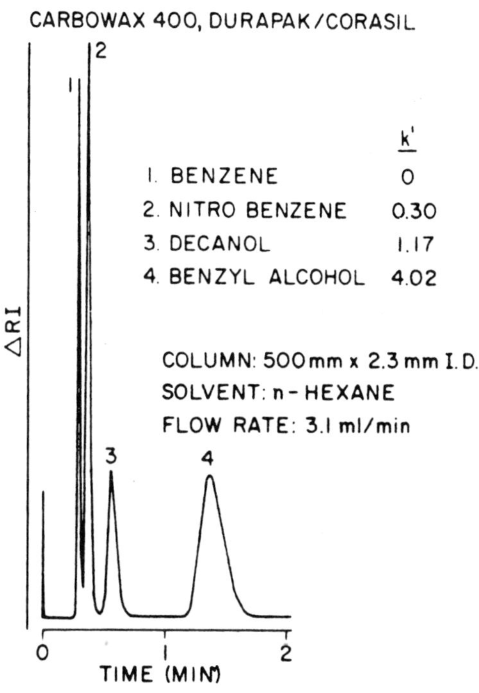

Figure 4.10. Separation of test mixture on column.
500 mm x 2.3 mm ID,
DURAPAK Carbowax 400/Corasil
Flow: 3.1 ml/min
Solvent: n-hexane
(Reprinted with permission from J. Chromatographic Science.)

Figure 4.11 shows a plot of efficiency (h) *vs.* linear velocity (u) for CORASIL based supports. The k' (capacity factor) values for the test compounds are approximately the same on both the conventionally coated and bonded supports. The efficiency at any given velocity is much better on the bonded support

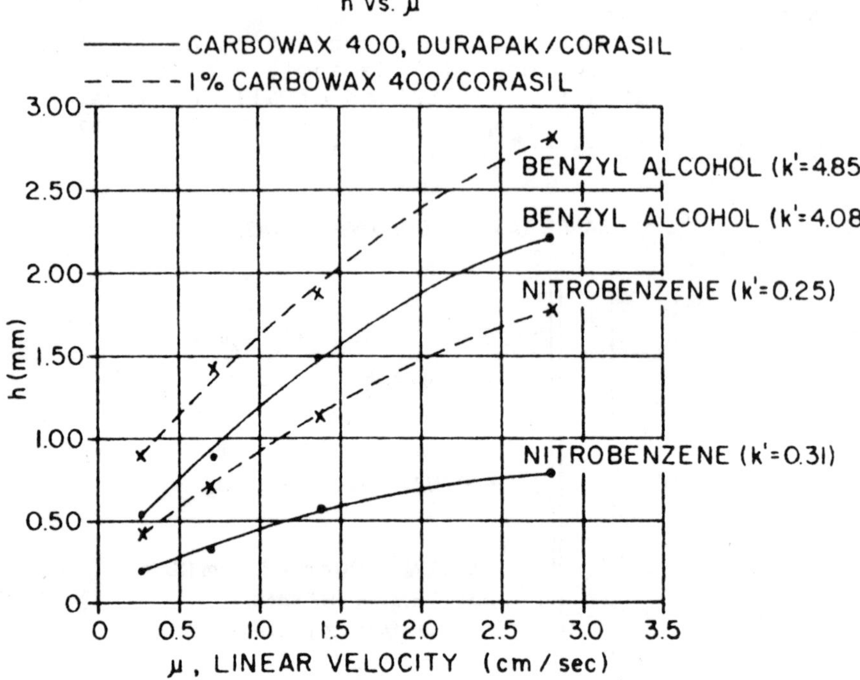

Figure 4.11. Plot of plate height vs. linear velocity. Solvent: n-hexane DURAPAK, Carbowax 400/Corasil (———) .1% Carbowax 400/Corasil (-----). (Reprinted with permission from J. Chromatographic Science.)

than on the coated support. This may be due to the fact that the liquid phase, when bonded, is more evenly distributed on the support, thereby eliminating "pooling" and consequently increasing the rate at which mass transfer takes place. These differences may also be due to the inability to pack columns as well when a liquid is coating the outside of the particles. Figure 4.12 shows the same result when the fully porous PORASIL was used as the base support.

Figure 4.13 compares the polarity of different molecular weight Carbowaxes bonded onto CORASIL. In this case, the most polar support has the lowest molecular weight bonded stationary phase, diethylene

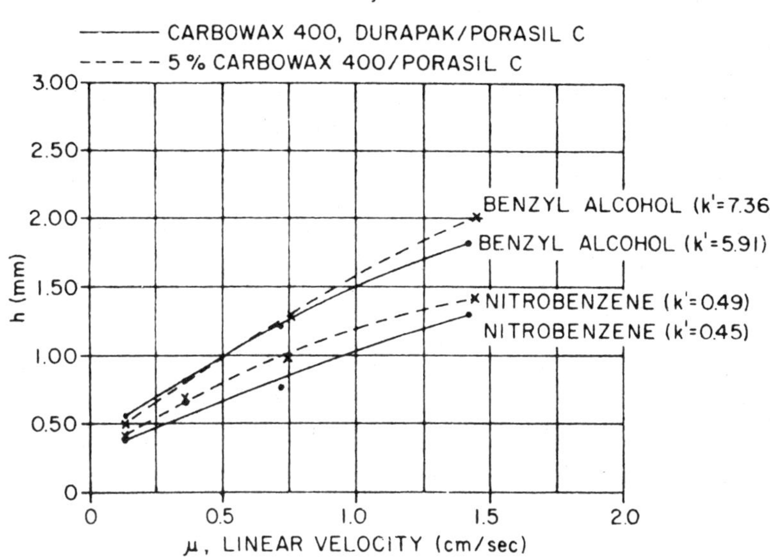

Figure 4.12. Plot of plate height vs. linear velocity. Solvent: n-hexane. Durapak, Carbowax 400/Porasil C (______). 5% Carbowax 400/Porasil C (-----). (Reprinted with permission from J. Chromatographic Science.)

Solvent: n-Hexane
Linear Velocity: 1.50 cm/ sec

Packings	Capacity Factor (k') Benzene	Nitro-Benzene	Decanol	Benzyl Alcohol
Carbowax 400/ Corasil	0	.26	1.08	3.95
Carbowax 4000/ Corasil	0	.21	.81	3.46
Diethylene Glycol/Corasil	0	.40	5.60	11.67

Figure 4.13. Effect of "brush length" on capacity factor (k'). (Reprinted with permission from J. Chromatographic Science.)

glycol, and the polarity of the packings decreases as the molecular weight of the Carbowaxes increases. This is the more expected order but just the opposite of that found in gas chromatography (refer to Figure 4.7).

The Carbowax 400 bonded supports have found greatest use in the separation of organometallic compounds. With some of these compounds, "plain" silica as a support is too harsh and causes either a decomposition or rearrangement of the compounds. Reverse-phase chromatography using polar solvents (water, methanol, etc.) many times reacts with the organometallic compounds. The intermediate polarity Carbowax 400 support used with intermediate and nonpolar solvents (chloroform, heptane, etc.) appears to be the right choice for selectivity and stability. Figure 4.14 shows the separation of the Cr, Zn and Cu diethyl dithiocarbamates and Figure 4.15 shows the separation of 2,3-Dimethylnaphthalene tri-carbonyl-chromium isomers.

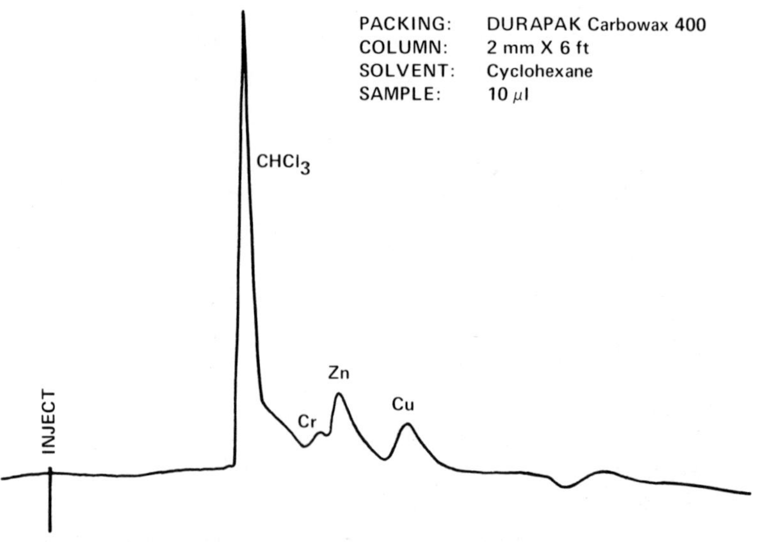

Figure 4.14. Separation of Cr, Zn and Cu diethyl dithiocarbamates.

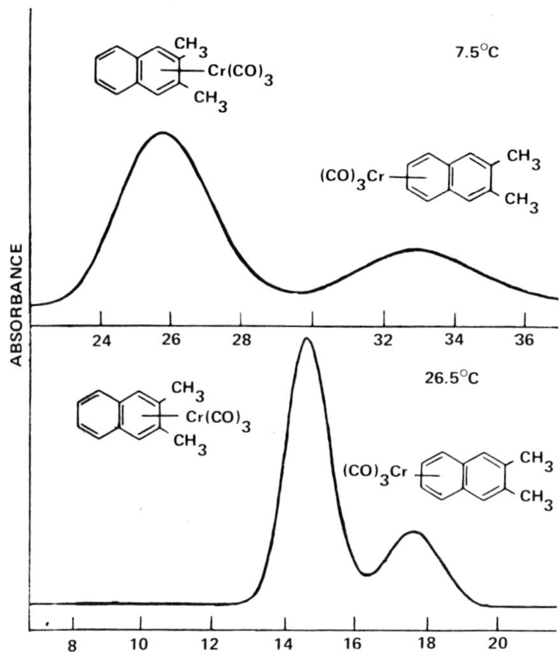

Figure 4.15. Liquid chromatography separation of chromium isomers. (Reprinted with permission from J. Organometal. Chem.)

Another phase of this work on bonded supports involved the bonding of organosilanes onto silica. Figure 4.16 shows the type of reaction involved and the stationary phases bonded onto silica. In the case of the two hydrocarbon phases, C_{18} and phenyl, an analytical (CORASIL) and a preparative version (PORASIL) were prepared.

In trying to understand the selectivity differences between the two hydrocarbon phases, C_{18} and phenyl, several test compounds were evaluated. It was initially thought that polynuclear aromatic compounds should separate best on the phenyl packing.

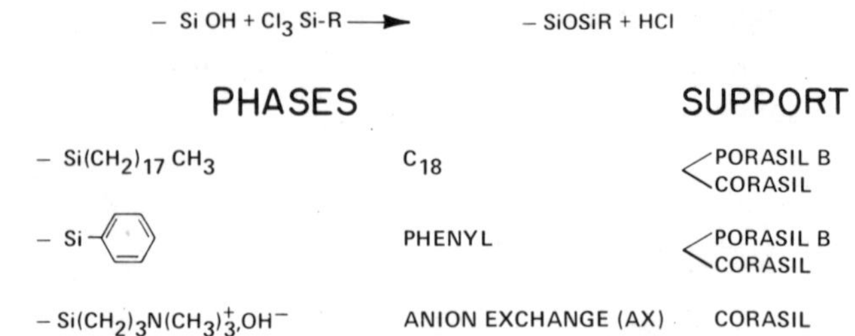

Figure 4.16. Summary of liquid chromatography packings (silane bonds).

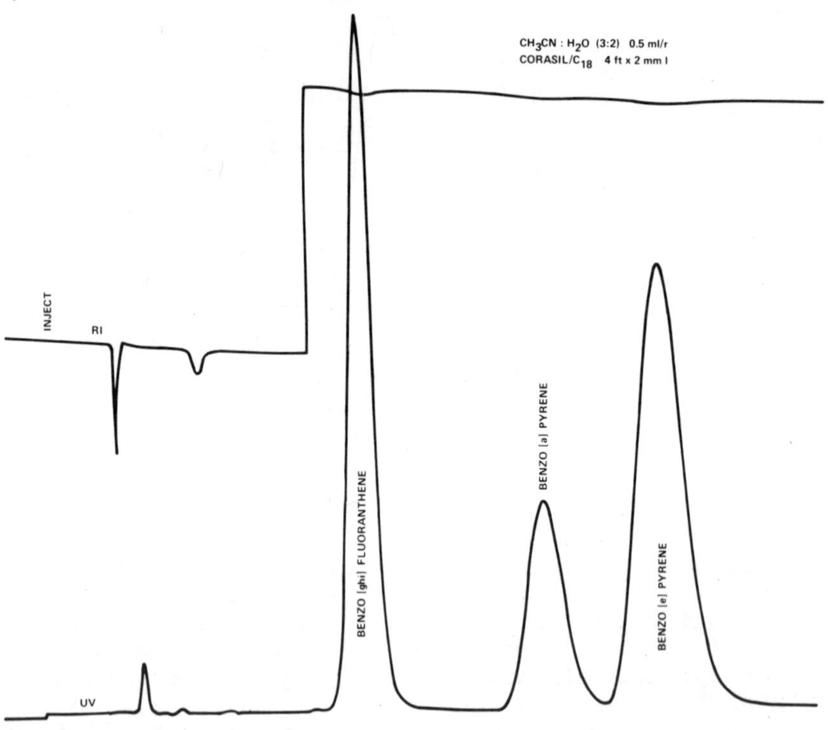

Figure 4.17. Separation of polynuclear aromatics
Stationary phase: Bondapak C_{18}/CORASIL
Solvent: Acetonitrile/water (3/2)

Experimentation showed that the C_{18} support gave better selectivity and peak shape for polynuclear aromatic hydrocarbons (Figure 4.17).

Figure 4.18 shows the separation of ascorbic acid (Vitamin C) on the anion exchange bonded support. This support behaves similarly to conventional ion-exchange supports. Figure 4.19 shows a separation of 5'-nucleotides using gradient elution (pH 3.0 to pH 4.3) on this support, BONDAPAK AX/CORASIL. This support can also be used with non-buffered solvents and "behaves" as a very polar bonded phase. Figure 4.20 shows the separation of several simple

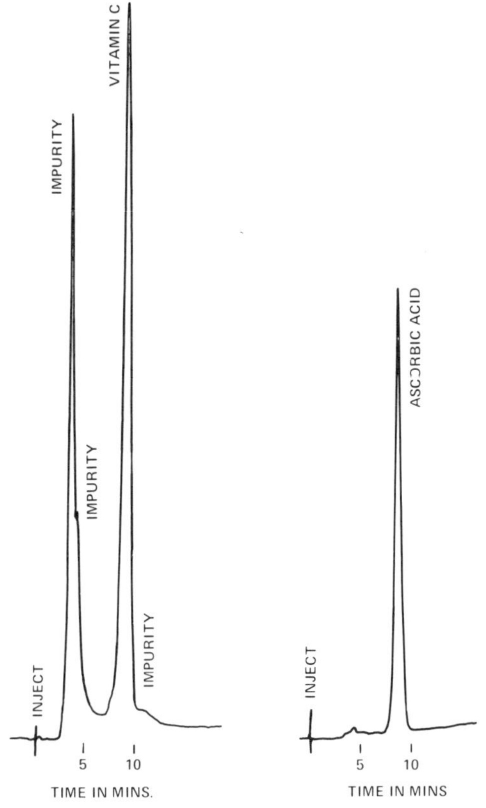

Figure 4.18. Separation of Vitamin C by ion exchange.

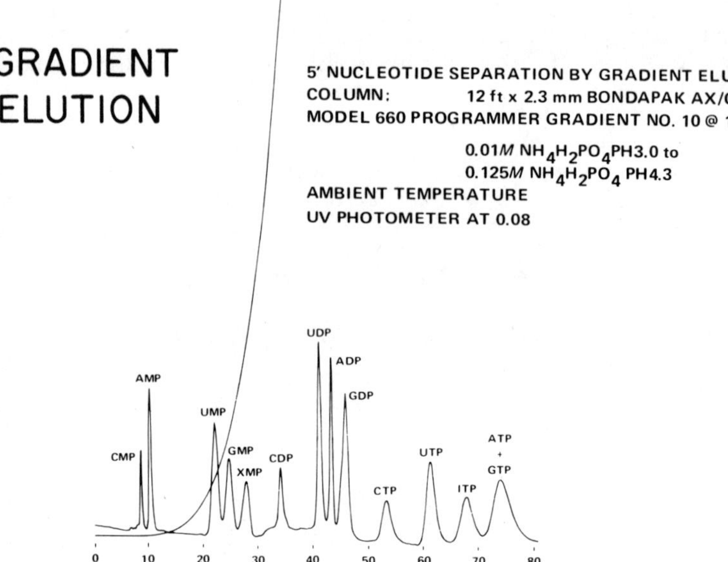

Figure 4.19. Separation of 5'-nucleotides by ion exchange and gradient elution.

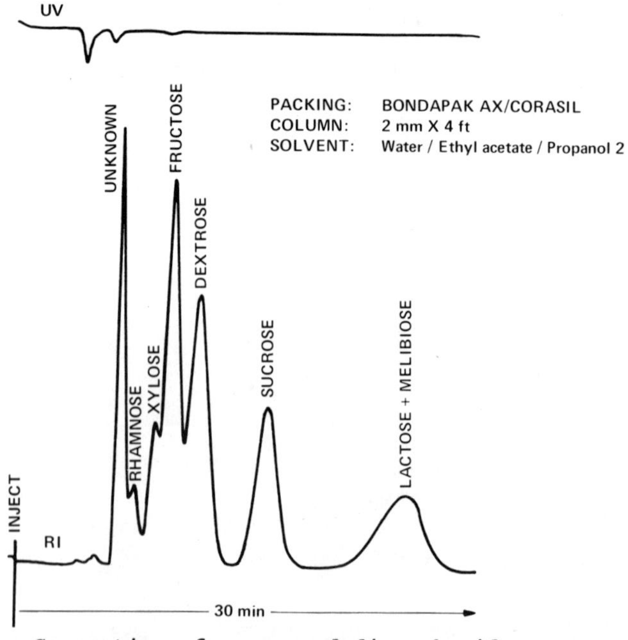

Figure 4.20. Separation of mono- and disaccharides.

saccharides on AX/CORASIL using a solvent of water, ethyl acetate and isopropanol.

Earlier, in Figure 4.16, it was shown that an analytical and a preparative version of C_{18} and phenyl supports were prepared. Figure 4.21 shows a preparative separation of two steroids on a 2' by 3/8" OD column packed with BONDAPAK C_{18}/PORASIL. This preparative packing has approximately fifteen per cent by weight of the C_{18} bonded to silica. This high percentage of liquid phase allows the large sample loads (272 mg) as shown in Figure 4.21 to be resolved without overloading.

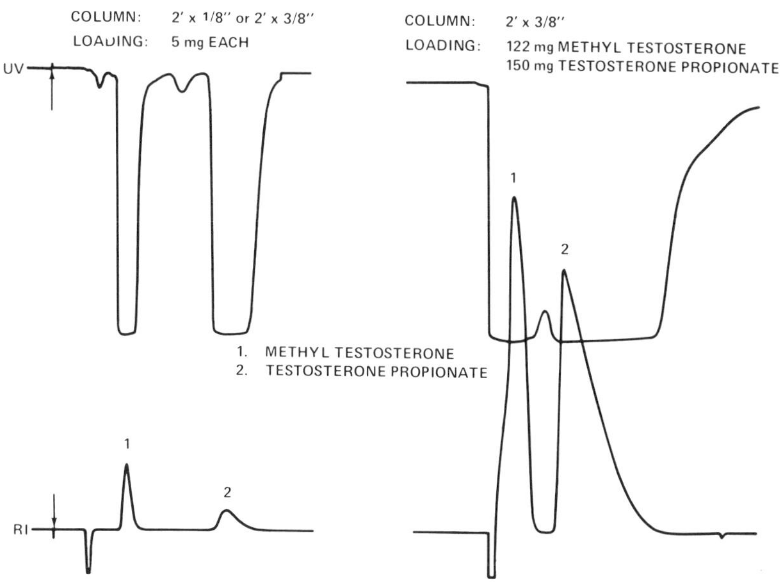

Figure 4.21. Preparative separation of steroids.

In trying to determine differences in the two hydrocarbon phases, C_{18} and phenyl, it was discovered that polar samples which were chromatographed by reverse phase gave much better selectivity and peak shape on the phenyl packing compared to the C_{18} packing. Figure 4.22 shows the separation of the

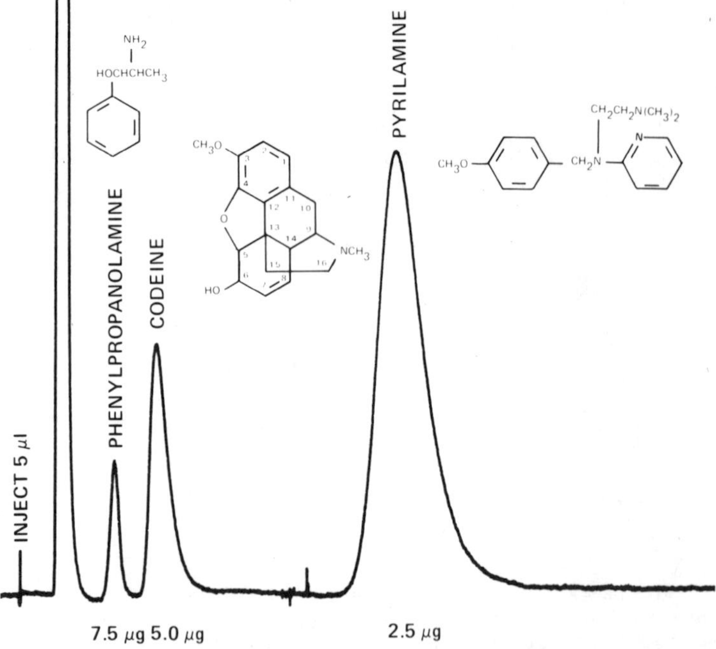

Figure 4.22. Separation of the active ingredients in a cough preparation.

polar active ingredients in a cough preparation using BONDAPAK Phenyl/CORASIL. Using C_{18}/CORASIL, the peaks are very skewed and do not lend themselves to quantitation. Figure 4.23 summarizes how this laboratory analyzes polar samples. If reverse phase is chosen, BONDAPAK Phenyl is used as the support. If normal phase is chosen, DURAPAK Carbowax 400 using acetonitrile solvent is used.

POLAR SAMPLES

	SOLVENT	SUPPORT	MODE
A)	METHANOL/WATER	PHENYL	REVERSE PHASE
B)	ACETONITRILE	CW400	NORMAL PHASE

Figure 4.23. Conditions suited for analyzing polar samples.

Figure 4.24 summarizes the LC bonded phases described in this chapter and lists them in order of polarity with the most polar support at the top.

POLARITY OF BONDED PHASES

$-\,Si\,(CH_2)_3\,N(CH_3)_3^+, OH^-$	AX
$O\,CH_2CH_2C \equiv N$	OPN
$-\,O\,(CH_2CH_2O)_X\,H$	CW 400
$-\,Si-C_6H_5$ 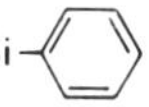	PHENYL
$-\,Si\,(CH_2)_{17}\,CH_3$	C_{18}

Figure 4.24. Polarity of bonded LC packings.

Further work on bonded phases is underway in this laboratory using silicas of 5 and 10 microns as the support and new stationary phases employing other electron donor and acceptor groups and phases containing optical activity.

REFERENCES

1. Little, J. N., W. A. Dark, P. W. Farlinger, and K. J. Bombaugh. *J. Chromatogr. Sci.*, *8*, 647 (1970).
2. Little, J. N., D. F. Horgan and K. J. Bombaugh. *J. Chromatogr. Sci.*, *8*, 626 (1970).
3. Horgan, D. F. and J. N. Little. *J. Chromatogr. Sci.*, *10*, 76 (1972).

CHAPTER 5

PELLICULAR ION EXCHANGERS

Csaba Horvath

INTRODUCTION

Ion exchange chromatography is probably the most prominent branch of liquid chromatography and is widely used for the separation of water-soluble substances which contain ionizable functions. The technique is called ion exchange chromatography because the stationary phase is a solid porous matrix with fixed ionogenic groups, most commonly an ion exchange resin. The interaction between the solutes and the stationary phase usually involves several physicochemical phenomena, and "the mechanism of the separation may be as far removed from exchange ions as chromatography is from color" as stated by Helfferich.[1]

The use of ion exchange resins in analytical chemistry[2] rapidly followed their commercial introduction. Early chromatographic applications[3,4] demonstrated the efficacy and versatility of ion exchange columns which have become indispensable separating tools in the laboratory, particularly in the life sciences.

Modern liquid chromatography began with the instrumentation of ion exchange chromatography. The amino acid analyzer which evolved from the fundamental work of Moore and Stein[5] was the first and is still the most widely used liquid chromatograph. Ion exchange chromatography of nucleic acids in a semi-automated fashion[6,7] was another important step toward the development of modern liquid chromatography.

Since then instrumentation has rapidly advanced in other areas of liquid chromatography, and today the technique approaches the speed, efficiency and convenience of gas chromatography. In order to distinguish the evolved chromatographic technique from conventional column chromatography, the term *high performance liquid chromatography* (HPLC) is frequently used.

The acronym HPLC can stand also for "high pressure liquid chromatography," since the technique employs relatively high column inlet pressures unlike traditional liquid chromatography. Those who find that the price of the instrumentation and column materials is high with respect to the cost involved in conventional column chromatography are also gratified by the acronym HPLC, because it may equally stand for "high priced liquid chromatography." Nevertheless the gain in separation efficiency, sensitivity and reproducibility, the drastic reduction in the time required for a chromatographic run, and the versatility of the instrumentation suggest that in most instances HPLC is a more economic analytical tool than classical liquid chromatography.

It appears that three factors are mainly responsible for the emergence of HPLC.

1. the application of the broad theoretical and practical experience gained in gas chromatography to the design of liquid chromatographic systems in general
2. the development of novel stationary phases for high efficiency columns which can be used at relatively high flow velocities and thus yield rapid separations
3. the availability of sophisticated instrumentation which entails various highly sensitive detectors, precise and reliable pumping systems for isocratic and gradient elution at high pressures, as well as appropriate temperature control and sampling devices.

In this chapter a family of novel ion exchangers, which have been developed especially for HPLC, is discussed. The unique feature of these column materials is that the actual stationary phase is deposited on the surface of fluid impervious spherules such as glass microbeads. In order to distinguish such shell-structured ion exchangers from conventional ion exchange resins, they are called pellicular or peripheral ion exchangers. Sometimes the terms solid core or superficial ion exchanger are used in the literature. The properties and chromatographic application of pellicular ion exchangers have already been treated in recent reviews on this subject.[8,9]

CLASSIFICATION OF ION EXCHANGERS

In most applications ion exchangers are used in beds packed with spherical or irregularly shaped particles whose mass entirely consists of an ion exchange resin. Such conventional ion exchangers are available in great variety with respect to the nature of the ionogenic groups and the matrix as well as the particle size.[10,11] Numerous methods have been developed for the synthesis of such resins and are well documented in the literature.[12,13]

Although inorganic materials such as circonium phosphate and hydroxy apatite are occasionally used as ion exchangers,[14] most conventional ion exchange resins have an organic matrix—usually a cross-linked polymer. They are divided into two major classes, cation and anion exchange resins, which contain negatively and positively charged fixed groups, respectively. Depending on the dissociation constant of the ionogenic groups of the resin we distinguish between weak and strong ion exchangers within both classes. Strong cation exchange resins usually contain sulfonic acid groups attached to the polymer matrix. Weak cation exchangers have fixed carboxylic groups. Strong and weak anion exchange resins have quaternary ammonium functions and weakly basic amino groups, respectively.

The most widely used strong ion exchange resins are made from styrene-divinylbenzene copolymer beads, which can be of microreticular (gel type) or macroreticular (macroporous type). The gel type ion exchangers become porous upon swelling and the pore size depends mainly upon the degree of cross-linking which is usually expressed by the per cent of cross-linking agent (divinylbenzene) in the copolymer. In chromatography such microreticular resins are employed almost exclusively. Macroreticular resins have a rigid macroporous structure because the copolymerization is carried out with a large amount of cross-linking agent in the presence of an inert diluent which is removed later in the process. These resins are less suitable for chromatography of small and medium-size solutes probably because the inhomogeneity of the pore size distribution of the micropores in the actual resinous phase results in poor efficiency.

Another important group of ion exchangers, used mainly in biochemical separations, is made by attaching ionogenic groups to an insoluble carbohydrate matrix. The diethylaminoethyl (DEAE) and carboxymethyl derivatives of cellulose are widely used weak anion and cation exchangers, respectively. Spherical cross-linked dextrane beads having similar functionalities also find broad application in ion exchange chromatography of biological substances.

Excellent books are available which deal with both the theory and practice of ion exchange[15] as well as with the analytical applications of ion exchangers.[16,17] Selected topics on ion exchange are treated in the chapters of a specialized series.[18,19] The use of ion exchangers in chromatography has also been the subject of numerous texts.[1,20,21] Since the physicochemical phenomena underlying ion exchange chromatography are the same for both conventional and pellicular ion exchangers, a great deal of the information available from the literature is pertinent to the present discussion and useful for the understanding and optimization of separations on pellicular ion exchange columns.

EVOLUTION AND STRUCTURE OF PELLICULAR ION EXCHANGERS

In most processes involving columns packed with conventional ion exchange resins, the rate-determining step is the solute diffusion inside the resin particles.[22,23] Since the diffusivity increases with decreasing degree of cross-linking, the intraparticulate mass transfer could be enhanced by lowering the resin cross-linking. The excessive swelling and shrinking as well as the poor mechanical properties of low cross-linked resins, however, make their use in packed columns impractical. Thus, the degree of cross-linking of commercial resins represents a practical compromise and is about 8% for polystyrene based resins. Nevertheless the diffusivity of most solutes of interest is at least two orders of magnitude lower in such resins than in the surrounding solution.

In order to meet the need of HPLC for column materials with rapid exchange characteristics, conventional resins in microspherical form have been introduced. The resins are available in a narrow particle size range with a mean particle diameter of 5 to 20 μm. Because of the relatively small particle size the diffusion path length of the solute is small in both the stationary and mobile phase and as a result high speed and efficiency can be obtained.[24,25] Since the whole mass of the conventional ion exchangers is made up of the actual resin, such columns have high loading capacity. On the other hand high inlet pressures are needed with columns packed with microspherical resins because the pressure drop necessary to maintain a given flow velocity through the column is inversely proportional to the squared particle diameter.

In the ion exchange chromatography of organic substances the intraparticular diffusion remains the controlling factor even with microspherical resins (C. D. Scott, private communication), and an improvement could be expected only by a further decrease in the particle diameter. The available column inlet pressure, the difficulties in obtaining uniform column packing and the instability of the bed at large pressure gradients, however, limit the minimum particle size for conventional resins. In addition the swelling and shrinking of such resins with changing ionic strength of the mobile phase can impose serious constraints on the permissible operating conditions.

The limitations of conventional ion exchange resins with respect to the slowness of the intraparticular diffusion were realized long before the advent of HPLC. Two basic concepts have been put forward to reduce the diffusion path length in the actual ion exchange resin independently from the particle size, thus departing from the conventional resin configuration.[26]

Superficial Ion Exchange Resins

The first nonconventional resin type is the superficial ion exchange resin which originates from the suggestion of Weiss[27] to convert only the outer shell of suitable polymer beads into ion exchange resin.

Parrish[28] prepared a superficial cation exchange resin having a capacity of 5.3 microequivalents per milliliter by sulfonating a shallow surface layer of cross-linked polystyrene beads for use in chromatography. Skafi and Lieser[29,30] carried out a detailed study on the preparation of such resins; recently Fricke *et al.*[31] also described a method for making superficial resins. The results of Skafi and Lieser indicated that the selectivity of their resin was similar to that of corresponding conventional resins, but only at a very low capacity of 0.6 per microequivalents/gram as compared to 5000 microequivalents per gram for conventional resins such as Dowex 50 x 12, was the diffusion in the resin shell not the rate determining step.

It appears that this approach to make useful resins for chromatography is beset by several problems. If the cross-linking of the polymer is low then the density of the sulfonic acid groups, and consequently the pore size of the swollen resin, gradually decreases toward the bead center. Since the solute diffusivity in the less swollen region can be orders of magnitude lower than in the highly swollen outer layer, the chromatographic efficiency which is determined by the slow diffusivity in the depth of the layer is poor. A uniform ion exchange resin layer and a sharp boundary between the resin and the unreacted polymer can be obtained if the sulfonation is truly a shell progressive reaction, but this normally occurs only if the cross-linking of the polymer is high. Then the swelling of the highly cross-linked resin layer and, consequently, the solute diffusivity inside the resin, however, are low and poor efficiency is obtained.

Although rapid sulfonation with a powerful sulfonating agent at high temperatures could at least theoretically yield shell progressive reaction with polymers of reasonably low cross-linking, this approach has not been described for making superficial ion exchangers for HPLC. Besides the above mentioned limitations the lack of wider interest in superficial resins can be explained also by the fact that the technique of preparing superficial resins by modifying the outer surface of polymer beads has been restricted to strong cation exchange resins.

Supported Ion Exchange Resins

The other fundamental type of nonconventional ion exchanger is the supported ion exchange resin which is obtained by forming a thin layer of resin in the cavities of an inert porous support.

Boardman[32,33] prepared such resins in a sealed tube by precipitation copolymerization of suitable monomers in methanol solution in the presence of Celite. Thus, the copolymer was deposited in the porous structure of the support material. A weak cation exchange resin was made from methacrylic acid and divinylbenzene, which had a capacity of

0.69 milliequivalent per gram. For the preparation of a strong cation exchanger, styrene and 5% divinylbenzene were copolymerized onto silanized Celite in a similar procedure and the product was subsequently sulfonated. The capacity of this resin was 0.28 milliequivalent per gram. The main reason for preparing such resins was to obtain higher resin capacities than those of conventional resins in the chromatography of proteins. Since large molecules cannot penetrate resins having the usual degree of cross-linking, the increased surface-to-volume ratio of the supported resins yielded higher protein adsorption values than those obtained with conventional resins.[34] As the support material can impart adequate mechanical stability to the resin, supported ion exchange resins can have a much lower degree of cross-linking than usual conventional resins and still be used in columns. Feitelson and Partridge[35] prepared a supported strong cation exchange resin with cross-linking of the order of 0.5 to 1.0%. The material was used for separation of large peptides by elution chromatography.

Supported ion exchange resins offer a greater column stability than liquid ion exchangers, which also have been used on porous supports,[36,37] since a completely insoluble lightly cross-linked resin layer is formed in the pores of the silanized support. Supported ion exchangers which could qualify as bonded phases have also been employed outside of column chromatography. Since aluminum hydroxide was first used to impart ion exchange properties to the paper,[38] a variety of ion exchangers has been used on paper support in paper chromatography. For instances, polyethylene imine impregnated cellulose, a supported anion exchanger, has found applications in thin-layer chromatography.[39]

With the development of various reactive organosilanes for use as adhesion promoters,[40] it has become possible to covalently bind an organic moiety to the surface of porous support via siloxane bonds. The treatment of silica with organosilanes already resulted in a variety of bonded phases for HPLC.[41,43] Since the siloxane bond between the organosilane and the silica surface is resistant to acidic aqueous media, suitable ion exchanger sorbents can be prepared for HPLC in this way if the pendant groups contain one or more ionogenic functions. These column materials can also be considered as supported ion exchange resins, which in the limit contain only a monomolecular layer of the actual ion exchanger. Although it is possible to cover the silica surface with such a molecular fur of the ion exchanger only, in most instances a very thin, slightly cross-linked anionic or cationic polysiloxane layer is bound to the surface.

A number of reactive organosilanes are available which either have ionogenic functions or allow the introduction of suitable ionic groups into the organic moiety after completion of the surface reaction. Taking into account that not only the nature of actual ion exchange functions but also the surface characteristics and the pore structure of the support can be varied, this approach can lead to a family of supported ion exchangers in both microparticulate and pellicular form for use in HPLC. Even if the ion exchange capacity per unit volume of such column materials is smaller than that of the conventional resins, the macroporous structure which facilitates intraparticular diffusion, and the favorable physical configuration of the actual ion exchanger promise to confer superior chromatographic properties.

In these ion exchangers the accessibility of the ionic sites to the solute molecules is expected to be more uniform than in macroreticular resins. With a support having a very uniform pore structure of the proper dimensions, a rigid mechanically stable macroporous ion exchange sorbent could be prepared in which the fixed ionic groups are—at least theoretically—equiaccessible to solute molecules having a given size range.

It is noted that even gel type ion exchange resins show considerable heterogeneity toward larger solutes due to the nonuniform structure of the cross-linked polymer which manifests itself in tailing in chromatography. Microreticular ion exchange resins, however, were developed for the exchange of relatively small ions and their employment in the chromatography of organic substances has been secondary. It is safe to assume that high chromatographic performance in the separation of large molecular weight solutes requires ion exchangers which are made primarily for this application.

There is no reason why there could not be porous supports other than silica also converted into ion exchange sorbents. Porous glass, zirconia, alumina and other materials can be treated with reactive organosilanes so that in the product the organic pendant groups are attached to the surface via stable covalent bonds and serve as anchors for the ionic functions. Obviously the chromatographic properties of the product are largely determined by the particular pore geometry, the surface coverage, and the properties of the organic moiety as well as by the nature of the fixed ionic groups.

Pellicular Ion Exchangers

The concept underlying the superficial ion exchange resins has been successfully reduced to practice by forming a thin ion exchange resin layer on the surface of glass microbeads, so that it cannot peel off in the column even if it is subjected to drastic elution conditions.[44,45] The advantages of such pellicular resins over superficial resins are obvious:

1. the resin shell can be made uniformly with any desired thickness
2. there are no limitations regarding the chemical composition and the cross-linking of the resin
3. uniform and spherical glass beads are easily available in any desired particle size range
4. due to the mechanical stability, chemical inertness and high density of glass stable columns are obtained and the packing of the column with pellicular materials is greatly facilitated.

The first pellicular ion exchangers were made with polystyrene-based strong anion and cation exchange resins. They can be used as general purpose ion exchangers in the same way as the corresponding conventional resins. Since then a variety of similar products with different resin structures and shell thicknesses have been commercially available (Reeve Angel). For instance glass beads coated with a weak aliphatic anion exchange resin or porous nylon are marketed for specific chromatographic applications.

A second type of pellicular ion exchange resins represents the combination of the concepts of superficial and supported ion exchange resins. Glass beads are first coated with a relatively inert porous layer and the ion exchange resin is deposited in the relatively large pores (80-200 nanometer) of the "controlled surface porosity support."[46,47] These pellicular resins could be called more precisely "peripheral supported ion exchange resins" in order to express the fact that actual ion exchange resin is supported in a peripheral porous layer on glass beads. A strong aliphatic cation exchanger as well as a weak aliphatic and a strong polysiloxane type anion exchanger of this type are commercially available (duPont).

Such resins embody the above mentioned advantages of the pellicular configuration. Since only a fraction of the outer shell of the particles consists of the actual ion exchange resin, the loading capacity of the column is further reduced. The column efficiency, however, is increased because of the higher surface-to-volume ratio of the resin, which is typical for supported resins. As a result of the high resin-support interface, this type of resin may not withstand such a wide range of operating conditions as the first type of pellicular resins since a breakdown of the resin is more likely to occur. The stability, however, can significantly be improved by covalent bonding of the resin to the siliceous support in the case of polysiloxane ion exchangers.

When pellicular silica of relatively high specific surface area is converted by silane treatment into an ion exchanger sorbent, the third type of pellicular ion exchangers is obtained. Such materials could be more precisely termed peripheral ion exchanger sorbents.

As the actual ion exchange functions are confined to the internal surface of the porous sorbent, the bulk properties of the ion exchange resin vanish and the intraparticular diffusivity is determined mainly by the pore structure of the sorbent. Relatively little is known about the actual structure of this type of pellicular ion exchangers which have been recently introduced commercially (Waters Associates, EM Laboratories). The macroporous structure of the ion exchanger sorbent shell facilitates intraparticular diffusivity and reduces the effect of solute-matrix interactions. Therefore, elevated column temperatures, which are often required to facilitate mass transfer inside the resin, may not be necessary. The overall chromatographic behavior and the stability of such pellicular ion exchangers in a wide range of elution conditions have yet to be investigated, and the structure as well as the ion exchange and sorption properties characterized.

The conceptual evolution of the different pellicular ion exchangers is schematically illustrated in Figure 5.1. The departure from the conventional resin configuration leads to superficial and supported resins, both having a relatively short characteristic

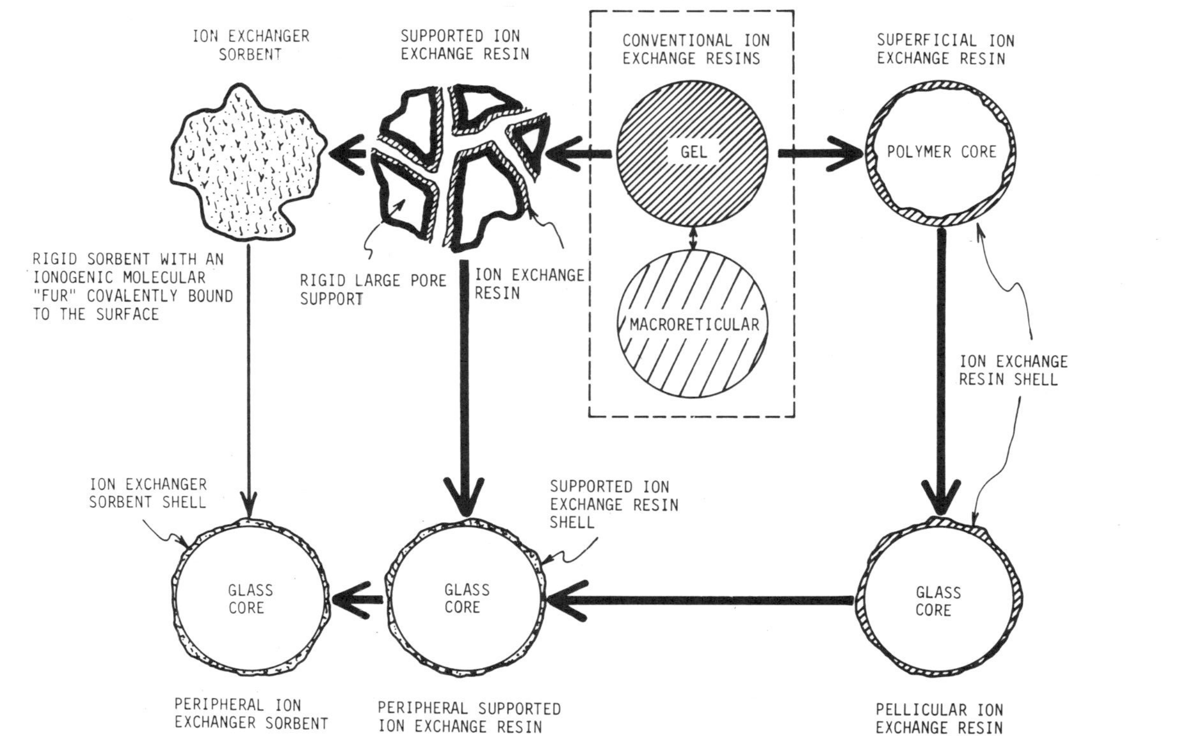

Figure 5.1. Conceptual evolution of the various pellicular ion exchangers used in high performance liquid chromatography.

length of the active medium which is independent of the particle size. Shell structured conventional resins on glass beads, the first pellicular ion exchangers, represent a practicable application of the superficial resin concept. Supported resins in pellicular form yield the peripheral supported ion exchange resins. In the limit, when the resin layer in the porous matrix is so thin that no bulk ion exchange resin is present, the material becomes an ion exchange sorbent, that is, a rigid sorbent with an ionogenic molecular fur covalently bound to the surface. The pellicular configuration of such ion exchangers is denoted peripheral ion exchange sorbent.

METHODS OF PREPARATION

Whereas the preparation of conventional ion exchange resins is well documented, very little has been published on the manufacturing of pellicular ion exchangers. Even the composition of the active shell is only vaguely described in the manufacturers' literature. It is probably due to the fact that these products have only recently been commercialized and the processes involved are proprietary.

The formation of a well defined and stable resin layer on the surface of tiny spherules requires a complex procedure, and a great deal of experience is required to obtain reproducible products with respect to the thickness, chemical composition and pore structure of the ion exchanger shell. Due to the difficulties involved in the particular topochemistry, however, commercial products frequently show batch to batch variations, but the situation is expected to improve. Indeed it is essential that a chromatographic result of the investigator does not remain an individual scientific curiosity.

The starting material is usually glass beads which are first classified, freed from metal particles and washed. The particle size is most commonly in the 30 to 40 μm range. Gross deviations from the uniform spherical shape can influence the outcome of the further treatment and have deleterious effects on the quality of the final product. For the preparation of ion exchangers the use of soda lime glass is preferred because of its relatively high resistance to aqueous media.

The formation of the chromatographically active shell on the surface of the beads can be carried out in many ways, depending on the particular type of the pellicular ion exchanger to be prepared. Some of the methods which have yielded useful products in the author's laboratory (even if the procedures have not been refined to obtain batch to batch reproducibility) are outlined as follows.

Formation of a Resin Layer on the Glass Surface

When fine glass beads, whose surface has been pretreated to procure satisfactory adherence, are coated with the monomer mixture, it is possible to carry out the polymerization so that a resin layer is formed *in situ* on the surface of the inert core. The surface treatment of glass is carried out with a suitable adhesion promoter[40] which preferably has polymerizable functions to establish covalent links between the glass surface and the polymer.

In one process, surface-treated glass beads are first coated with a mixture of styrene, divinylbenzene and an initiator. Then the polymerization is carried out in an aqueous slurry like the suspension polymerization process used in the manufacture of conventional resins. Alternatively, the monomers and initiator are dissolved in a solvent which does not swell the polymer. The glass beads are suspended in the solution and polymerization is carried out under vigorous stirring so that the polymer precipitates onto the surface of the beads. This precipitation copolymerization process has a greater flexibility but gives a less well defined and more inhomogeneous resin coating than suspension polymerization. The control of the thickness of the polymer layer formed by these processes requires a very precise adjustment of the conditions such as the stirring rate, temperature and the composition of the reaction mixture.

In the case of a cross-linked polystyrene coating, strong pellicular cation and anion exchange resins are obtained by sulfonation and by chloromethylation following amination with a tertiary amine, respectively. Although the chemistry involved in performing these reactions is essentially the same as used in making conventional resins, usually less severe conditions are needed to convert the thin polymer layer into an ion exchange resin. On the other hand, stirring has to be controlled in order to avoid the attrition of the polymer.

The resin layer can also be formed by colloidal particles in a fashion similar to the preparation of ion exchange membranes using finely ground resins.[48] Actually, pellicular resins can be considered as thin spherical ion exchange membranes supported by glass beads; thus, concepts used in the preparation of membranes are also applicable to their preparation. In order to obtain a pellicular resin, glass beads are coated with a paste or slurry containing the appropriate resin powder and a suitable binder such as colloidon in a volatile solvent. Then, the solvent is evaporated while the beads are agitated to avoid agglomeration.

A refinement of this approach has been described by Kirkland.[49] The resin coating on glass beads is formed by layers of fine ion exchange resin particles which alternate with interlayers of still smaller particles of silica. The coating is held together by opposite electrical charges of the adjacent layers. For example, an anion exchange resin can be prepared by treatment of glass beads, first with a suspension of Ludox AM, a very fine silica modified with alumina. The product is dried after each treatment and the procedure is repeated until a sufficiently thick layer is built up. The alternating layers of the positively charged resin and the negatively charged silica particles form a uniform and coherent coating. Essentially the same technique was described for making a pellicular strong cation exchange resin.[51]

In Situ Resin Formation in the Cavities of a Porous Surface Layer

For the preparation of peripheral supported ion exchange resins, first a porous shell has to be formed on the glass surface. Although methods have been described to form a porous layer of an inert support on the surface of glass beads for gas chromatography, the development of a novel technique for the preparation of a "controlled surface porosity support"[50] has been greatly advanced by HPLC. This commercially available pellicular support (duPont) was originally introduced for liquid-liquid chromatography and later used for the preparation of ion exchangers. The actual support shell is about one micrometer thick and consists of relatively inert siliceous particles in the 80 to 200 nanometer size range according to the manufacturer.

When the porous surface layer of this or similar material is impregnated with a suitable monomer mixture containing an initiator, a cross-linked polymer can be formed in the cavities of the layer *in situ.* Either an ionogenic monomer is used or the ionic groups are introduced into the polymer by subsequent chemical reaction. An advantage of this method is that polymerization can be carried out under a variety of conditions. For example, the impregnated beads are suspended in a liquid which is not a solvent of the components of the monomer mixture, and the polymerization is carried out by heating and stirring the suspension. The polymerization may also be carried out in a sealed container without a suspending liquid by heating or irradiation. In a similar fashion the porous layer can be impregnated with a soluble polymer which is then crosslinked *in situ.* With the "controlled surface porosity support" a strong cation exchanger has been prepared with a fluorocarbon polymer containing sulfonic acid groups.[47] A pellicular anion exchanger was obtained by using an acrylic polymer with suitable amino functions.[46] Later a polysiloxane resin containing quaternary ammonium groups was deposited in the cavities of the support so that concurrently the ion exchange resin was covalently bound via siloxane bridges to the surface of the porous siliceous support.[52] As a result the stability of the product was significantly enhanced.

Preparation of Peripheral Ion Exchange Sorbents

A variety of silica gel coated glass beads are commercially available for use as column materials in HPLC. The specific surface area of these stationary phases ranges from 8 to 15 square meters per gram of beads. The conversion to ion exchange sorbents can be performed essentially by silanization techniques similar to those employed in the preparation of bonded phases.[41,43] A weak anion exchanger can be obtained for instance by reacting the pellicular silica with γ-aminopropyltrimethoxysilane. By quaternarizing the amino functions a strong anion exchanger can be prepared.

Since a wide variety of trialkoxy- or trichloroorganosilanes are commercially available, various reactive organosilanes having suitable ionic groups can be readily prepared and then reacted with the sorbent. Alternatively the surface is first reacted with the silane, then ionogenic functions are introduced into the covalently bound organic moiety. This approach can be extended to other pellicular sorbents whose shells consist of zirconia, titania or alumina with a uniform pore geometry.

CHARACTERIZATION

Although methods for testing conventional ion exchange resins are extensively documented in the literature, no standard procedures have been described for the characterization of pellicular ion exchangers. So far their quality is solely judged by their chromatographic performance and stability. Despite the diversity of products whose chemical composition is not known in detail, there are several test procedures which can be employed to assess their quality besides the chromatographic evaluation.

Microscopic Examination

Visual observation of the beads under the microscope is the simplest method for the measurement of the sphericity and particle size distribution[53,54] as well as for the estimation of the thickness and uniformity of the ion exchanger coating. If there are undesired fines or metal particles in the product, they also can be seen by such an examination.

In order to make the ion exchange layer visible the beads are preferably stained. Although a number of dyes have been described for staining ion exchange resins,[55] acid fuchsin and Rhodamine B often give satisfactory results with anion and cation exchangers, respectively. First, the beads are immersed in the dye solution and then filtered and dried on air so that the moisture content is not completely removed. After placing the beads on a microscopic slide a drop of immersion oil is added. Thus, the glass core is almost invisible and the colored resin coating can be examined for uniformity and thickness under the microscope at substage illumination. Black spots in the beads, which are often observed under such conditions, are caused by air bubbles included in the glass. For a given type of resin the strength of staining is related to the layer thickness and the ion exchange capacity.

When the ion exchanger is anchored in a porous sorbent shell the observed colored layer can be misleading since the sorbent itself can also be stained in the absence of the actual ion exchanger. For example the staining of a siliceous cation exchanger with Rhodamine B does not necessarily reveal the presence of strong acid groups because silica gel takes up the dye also.

Sieve Analysis

The particle size distribution of the product is often determined by screening.[56] Although the aperture of the finest standard screen is 38 μm, micromesh screens are available which allow dry screening down to 20 μm particle diameter. Since spherical particles tend to clog the screen much more than irregularly shaped particles, frequent cleaning of the screens in an ultrasonic bath is necessary. The effectiveness of the screening procedure is also influenced by the moisture content of the material. Both high and low moisture content may lead to agglomeration due to a liquid film formation on the surface of the particles and to electrostatic charges, respectively. It is therefore recommended to condition the resin before screening by keeping it in an atmosphere of controlled humidity, for example, in a dessicator over salt solutions. Since free flowing resin is also needed when columns are packed in a dry procedure, the same treatment is recommended prior to the packing of columns.

Measurement of Ion Exchange Capacity

Although the ion exchange capacity is one of the most important specifications of conventional ion exchange resins, it appears to be a less significant property of pellicular ion exchangers. First, whereas these materials are employed in the separation of rather bulky organic molecules, the ion exchange capacity is traditionally measured with small sodium and chloride ions. Therefore, the chromatographic capacity and ion exchange capacity do not necessarily correlate. Second, adsorption of the solute on the resin also plays a very important role in the chromatography of organic substances on pellicular ion exchangers,[57] and the ion exchange capacity gives only limited information on this phenomenon. Consequently, the chromatographically relevant capacity of the ion exchanger could be best determined by an appropriate dye adsorption test colorimetrically. Although methylene blue is generally used for such measurements, optimum results would probably require the individual selection of the different dyes for the different kinds of ion exchangers.

Nevertheless the ion exchanger capacity of strong polystyrene-type pellicular ion exchange resins can be conveniently determined by potentiometric titration. Cation and anion exchange resins are first conditioned by acid and alkali washing, then converted into the hydrogen and hydroxyl form, respectively, and dried on air at room temperature. Strong cation exchangers can be titrated directly with sodium hydroxyde, but other resins are preferably titrated in salt solution. In most cases reproducible data are obtained at titration times of less than one hour. From the titration curve the ion exchange capacity is calculated as usual and expressed in microequivalents per gram of air-dry resin.

Titration curves can also reveal the presence of ionic groups of different strengths. For instance, pellicular polystyrene based strong anion exchangers consistently showed at least two different cationic groups in agreement with the observation made with conventional resins.[13] According to our experience the chromatographic properties of such products are usually determined not only by the total ion exchange capacity but also by the relative amounts of the weaker and stronger basic groups in the resin.

When the ion exchanger is on a siliceous support, the acidic silanol groups interfere with the titration and another method has to be found for the determination of the ion exchange capacity of the actual ion exchanger. For example, the sulfonic acid groups of a strong cation exchanger can be converted into a zinc salt since the silanol groups do not react. After removing the zinc ions by hydrochloric acid, the ion exchange capacity can be determined from the zinc concentration in the eluent, for example, by atomic absorption spectroscopy.[46] When the ion exchange capacity is very low the employment of radioactive tracers can be considered.[30]

A variety of polystyrene-based pellicular resins have been investigated by the titration method in which the measured ion exchange capacity ranged from 5 to 60 microequivalents per gram of dry beads, depending on the thickness and chemical composition of the shell. These ion exchange capacities are 200 and 1000 times smaller than those of conventional resins. For use in packed columns, however, it is more meaningful to compare the capacities of wet resins on a volume basis. Then the discrepancy is significantly smaller because of the relatively high density of pellicular resins, due to the heavy glass core, whose density is about 2.5 gram per cubic centimeter.

CHROMATOGRAPHIC PROPERTIES

Theoretical Considerations

The efficiency of an ion exchanger column is conveniently expressed by the plate height, H, which is a measure of band spreading per unit column length. Most generally it can be written as the sum of two plate height increments, which arise from the non-equilibrium phenomena caused by mobile and stationary phase processes; that is,

$$H = H_M + H_S \tag{1}$$

where H_M and H_S are the mobile and stationary phase plate height contributions, respectively. H_M depends on the geometry of the column packing, the detailed velocity profile of the eluent flow as well as on the diffusivity, D_M, of the solute in the mobile phase and the distribution ratio, k. In liquid chromatography, however, their exact relationship in determining the value of H_M is not known. On the other hand, H_S can be calculated for spherical ion exchange resin beads[58] as

$$H_S = \frac{1}{30}\frac{k}{(1+k)^2}\frac{d^2u}{D_S} \tag{2}$$

where d is the particle diameter, D_S is the effective solute diffusivity in the resin, and u is the average mobile phase flow velocity in the interstitial space. Therefore u is not the same as the chromatographic flow velocity that is measured by the retention time of an unsorbed solute.

By introducing the pertinent dimensionless groups such as the reduced stationary phase plate height increment, h_S, as

$$h_S = H_S/d \tag{3}$$

and the reduced velocity, ν, as

$$\nu = ud/D_M \tag{4}$$

we can express equation (2) in dimensionless form by

$$h_S = \frac{1}{30} \frac{k}{(1+k)^2} \frac{D_M}{D_S} \nu \quad = C_{sphere} \nu \tag{5}$$

C_{sphere} is the dimensionless chromatographic mass transfer parameter for spherical conventional ion exchange resins.

In order to compare the efficiency of a conventional to that of a pellicular ion exchanger we assume that the chromatographically active material has the same chemical and physical structure in both cases. In addition it is assumed that both are spherical, have the same particle size and yield the same packing geometry so that the value of h_M is invariant. Under such conditions the difference between the two column efficiencies is solely due to the different values of h_S.

The value of h_S can be calculated also for pellicular resins, and it has been shown[59] that with the above assumptions the expression is the same as that in equation (5); but the numerical factor 1/30 has to be replaced by the factor

$$\frac{1 - 5\ (1 - \delta)^3 + 9\ (1 - \delta)^5 - 5\ (1 - \delta)^6}{30\ [1 - (1 - \delta)^3]}$$

Here δ is the dimensionless shell thickness given by

$$\delta = (R_o - R_i)/R_o \tag{6}$$

where R_o and R_i are the outer and inner radii of the spherical annulus.

For sufficiently thin shells ($\delta < 0.1$) the configuration factor can be approximated by $5\delta^2$. Thus, for otherwise equivalent chromatographic conditions and at a fixed value of k the ratio of the mass transfer parameter of a conventional resin, C_{sphere}, to that of a pellicular resin, C_{shell}, is approximately given by

$$C_{sphere}/C_{shell} = 1/5\delta^2 \tag{7}$$

The ratio C_{sphere}/C_{shell}, which expresses how much greater is the value of h_S for conventional resins than for pellicular resins at the same capacity ratio, k, is plotted against the dimensionless layer thickness in Figure 5.2. It is seen that the pellicular configuration results in a great reduction of the

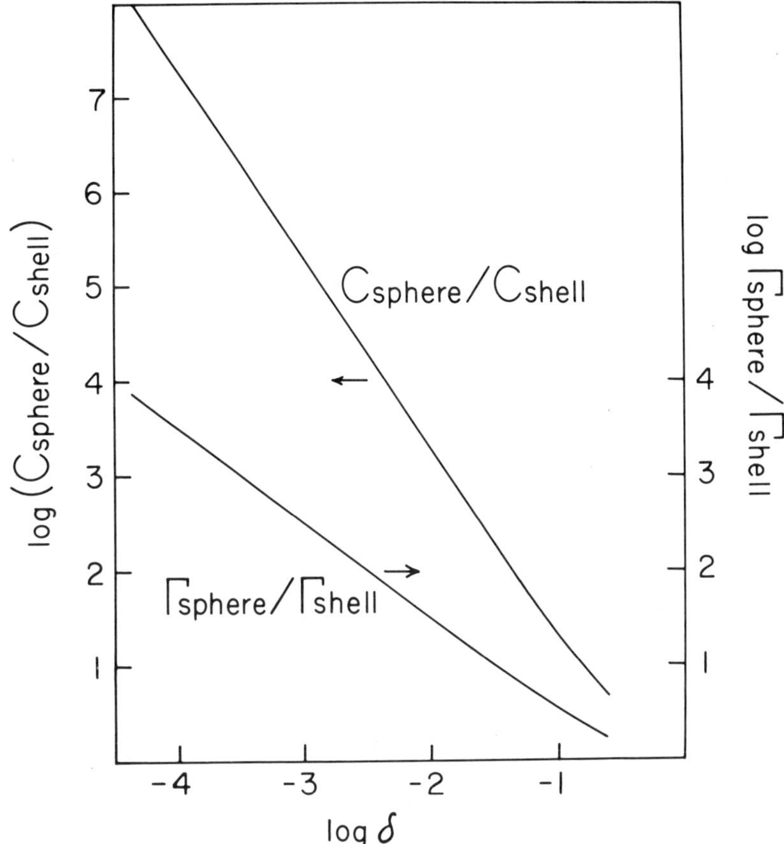

Figure 5.2. Comparison of conventional and pellicular resin beads. The ratio of the layer thickness to the particle radius, β, is 0.04 for the shell configuration. The volume of the small bead is equal to the shell volume.

stationary mass transfer contribution. For example, h_S can be reduced more than thousandfold by confining the resin to a surface layer having a thickness of 1 per cent of the particle radius.

Since only a fraction of the particle volume consists of the chromatographically active medium under such conditions as shown in Figure 5.3, the amount of stationary phase in a given column packed with a pellicular ion exchanger is smaller than if the column was packed with a conventional resin. At a given eluent strength the capacity ratio of a

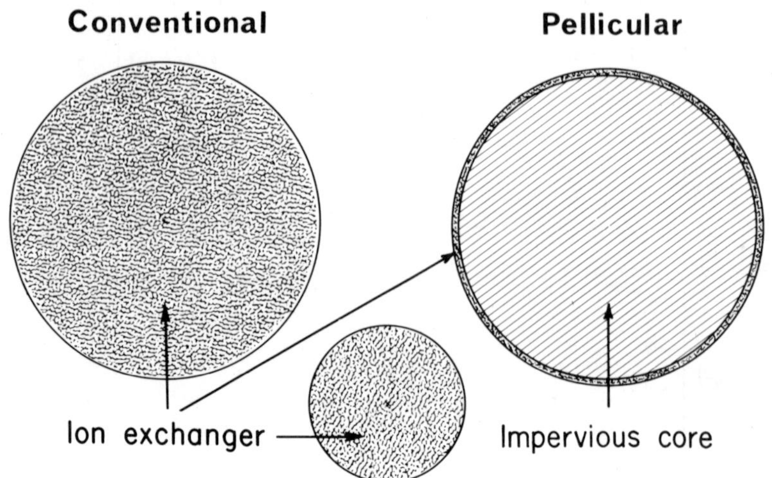

Figure 5.3. Plots of C_{sphere}/C_{shell} and $\Gamma_{sphere}/\Gamma_{shell}$ for equidiametral particles against the ratio of the shell thickness to the particle radius, β. In the calculation of C_{sphere}/C_{shell} the capacity factor, k, was assumed to be constant.[59]

solute depends on the thermodynamic equilibrium constant, K, and the so-called phase ratio, Γ, which is given by the ratio of the mobile phase volume to the stationary phase volume in the column, according to the following relationship:

$$k = K\,\Gamma \tag{8}$$

Therefore, the attenuating effect of the pellicular structure on the phase ratio is of great interest in chromatographic applications. The ratio of the phase ratios in a conventional resin column, Γ_{sphere}, to that in a pellicular resin column, Γ_{shell}, when the intraparticular porosity is the same in both columns, is given by

$$\Gamma_{sphere}/\Gamma_{shell} = \frac{1}{1 - (1 - \delta)^3} \tag{9}$$

where the denominator represents the volume fraction of the particle that is occupied by the resin shell of thickness, δ. The dependence of $\Gamma_{sphere}/\Gamma_{shell}$

on the layer thickness is also illustrated in Figure 5.2. It is seen that the decrease in h_S due to the reduction of the C_{shell} values by reducing the shell thickness is significantly greater than the corresponding decrease in the phase ratio.

This conclusion can also be inferred from the comparison of the three spherical particles in Figure 5.3. The two equidiametral beads represent conventional and pellicular resin beads, the latter having $\delta = 0.04$. The smaller bead has the same volume as the shell of the pellicular resin bead which equals 12 per cent of the volume of the large conventional resin bead. It shows that a given reduction of the diffusion path length represents a much lesser reduction in the actual volume of the resin. As the advantages of the pellicular structure are actually related to the high surface-to-volume ratios thus obtained, it is interesting to compare these values for the above beads. Assuming 100-micron diameter for the large beads, we obtain surface-to-volume ratios 60, 500 and 121 nm^{-1} for the large conventional, pellicular and small conventional beads, respectively.

Experimental results indicate that the capacity factors with pellicular sorbents are indeed lower by a factor of $[1 - (1 - \delta)^3]$ than those with conventional sorbents under identical elution conditions. Thus, the k values are roughly ten times lower when $\beta = 0.04$. The decrease in the column loading capacity is about the same as the decrease in the phase ratio.

In contradistinction, the improvement in column efficiency with pellicular sorbents is much less than would be expected from the reduction of the stationary phase plate height contribution alone. This can be easily explained by the fact that in liquid chromatography the plate height increment can be significant when aqueous eluents are used at the usual flow velocities and the solute molecules carry electric charges. Therefore advantages of the pellicular configuration can more realistically be assessed by plotting the ratio of the overall plate heights for conventional and pellicular particles, h_{sphere}/h_{shell}, against the ratio of the stationary phase plate height contribution for the conventional packing, h_S, to the mobile phase plate height increment, h_M. Figure 5.4 shows such plots for different values of the dimensionless shell thickness. It is seen that when about 90% of the plate height for the conventional packing is due to stationary phase effects, that is, $h_S/h_M \approx 10$, the improvement in the

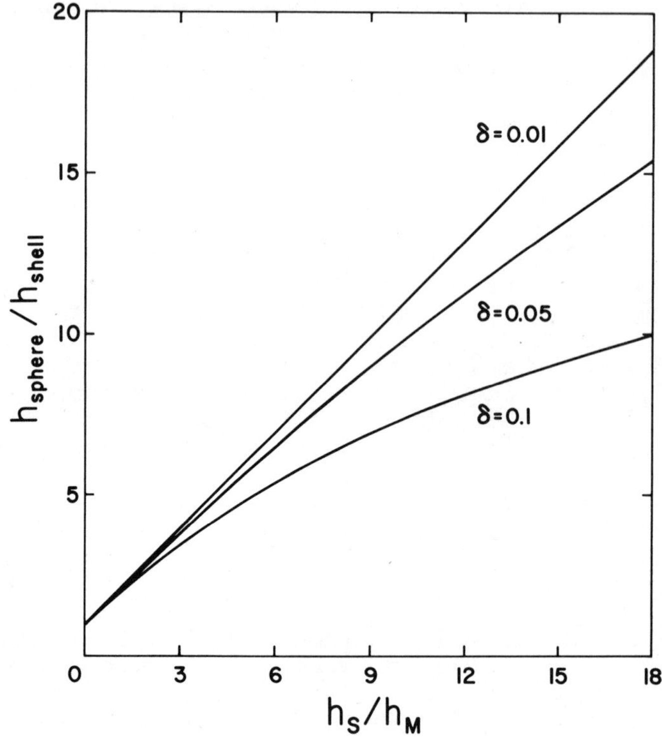

Figure 5.4. Plots of h_{sphere}/h_{shell} against the ratio of the stationary phase to mobile phase plate height contributions, h_S/h_M, for equidiametral particles with the dimensionless shell thickness, β, as the parameter.

column efficiency is about ten-fold with a pellicular sorbent having a dimensionless shell thickness of 0.05.

It has been shown[60] that the expression for the mobile phase plate height contribution is very complex even in the absence of electrical effects and interfacial mass transport resistances and given by

$$h_M = \left(\frac{1}{\alpha} + \frac{1}{\beta\nu^{\gamma}}\right)^{-1} + \omega\left(1 + \mu\frac{1}{1+k}\right)^2 \nu \qquad (10)$$

where α, β, γ, ω and μ are constants whose value has not been determined. They may vary with both the particle and column diameter as well as the packing technique since all of these can affect the configuration of the packing and the detailed flow profile of the eluent stream.

In view of the uncertainty regarding the constants in equation (10) it is difficult to predict the values of h_S/h_M for different packings and the potential improvement in column efficiency due to the pellicular structure. Nevertheless, h_S/h_M is expected to be significantly greater than unity in ion exchange chromatography under a wide range of operating conditions. Consequently, the column efficiency could always be increased by using pellicular instead of conventional ion exchangers, and micropellicular ion exchangers of 5 to 10 micrometer particle diameter would yield higher efficiencies than the conventional microspherical ion exchange resins of the same particle size in HPLC.

Practical Aspects

At present most column materials employed in HPLC can be divided into two classes: pellicular and microparticular. In ion exchange chromatography either conventional microspherical resins (about 15 μm in diameter) or pellicular ion exchangers, whose particle diameter is 30 to 50 μm, are used most frequently. Since the plate height decreases roughly with the squared particle diameter, small conventional resin particles give higher column efficiency than the relatively large pellicular resin particles. Consequently the column length can be drastically reduced and as a result fast separations can be obtained with very fine conventional resins. Due to the relatively high loading capacity, short column length and high efficiency, the solute concentration in the effluent can be high and trace analysis is facilitated.

Table 5.1 shows column performance data measured with nucleoside samples using conventional and pellicular cation exchange resin columns. The values for microspherical resin columns have been calculated from data of Burtis *et al.*[35] The conditions were optimized to achieve rapid separations; thus, the results can give insight into the behavior of these columns in practical applications. Conventional resins, which had five to ten times smaller particle diameter than the pellicular resin, gave higher column efficiencies as measured by the plate

Table 5.1

Column Performance with Conventional Microspherical and Pellicular Cation Exchange Resins in the Separation of Nucleosides

Resin	*Particle Size (μm)*	*Plate Height (mm)*	*Plates/Sec*	*Plates/Atm*
Conventional[a]	3-7	0.29	6.2	2.9
Conventional[b]	7-14	0.74	4.6	1.0
Pellicular	44-53	1.94	3.1	22.3

[a]DC-X resin, Durram Chemical Co.

[b]VC-10 resin, Sondell Scientific Instruments.

height. The plates per second values expressing the speed of separation are also somewhat higher with the microspherical resins. However, the plates per atmosphere values, which express the column inlet pressure required to obtain the stated plates per second, are an order of magnitude higher for conventional resins than for the pellicular resin. These observations are supported by theoretical considerations and suggest that, whereas both column types facilitate equally rapid separations, pellicular resin columns require less column inlet pressure.

In view of this comparison pellicular resins of relatively large particle diameter would have no net advantage over conventional microparticulate resins when the column inlet pressure is not a limiting factor. In practice, however, the pressure capability of the instrument is often limited and the use of pellicular resins is advantageous. In addition, several features of pellicular ion exchangers, which are not directly related to column efficiency, make them preferable in many chromatographic applications. These practical benefits of the pellicular structure arise from the combination of the actual ion exchanger with the heavy and rigid glass support, which imparts favorable handling characteristics and stability to the column material.

Columns can be packed with pellicular ion exchanger by using the dry packing procedure, which is rapidly performed without any special equipment, and the reproducibility of the packing is excellent.

In contradistinction, microparticulate resins require a cumbersome wet packing procedure at high pressures.[61] The packing of small bore columns, 0.5 to 1.0 mm i.d., which have low eluent consumption and are particularly suitable for the analysis of minute samples, is facilitated by pellicular ion exchangers.

The pellicular structure permits the use of stationary phases that have promising chromatographic properties but would not yield stable columns in conventional form. For instance, conventional resins with less than 8% cross-linking do not withstand the pressure gradients used in microparticular resin columns. On the other hand, resins with much lower degree of cross-linking can be prepared in stable pellicular form. Similarly this configuration of supported resins and ion exchange sorbents also gives stable columns at high pressures. The variety of novel ion exchangers, which have been introduced in pellicular form for use in HPLC, clearly illustrates that this configuration facilitates the preparation of stationary phases with superior chromatographic properties.

Unlike many conventional resins, pellicular ion exchangers are usually stable toward changes in eluent composition or column temperature and have long life even under greatly varying elution conditions. The reconditioning of pellicular resin columns can be performed rapidly at high flow rates; thus the overall analysis time is reduced.

The relatively low phase ratio in pellicular resin columns can be of advantage when otherwise strongly retarded solutes are separated, because they can be eluted without resorting to extreme eluent strength or temperature. Rapid separations can frequently be obtained by isocratic elution and at ambient temperature.

A major disadvantage of the pellicular resins, however, arises also from the relatively low phase ratio, which results in a low column loading capacity. Consequently the peak capacity is lower than with conventional ion exchangers. Although sensitive detectors are now available, trace analysis and the separation of samples, which contain more than 20 components, is usually not successful with the presently available pellicular ion exchangers having relatively large particle diameter and low phase ratio. Since the band spreading due to mobile phase effects is relatively high in such columns, the plate height is also significantly higher than that obtained with microparticulate resin columns. Nevertheless, the high column permeability permits the operation of long columns at relatively low inlet pressures.

Since the sorbtion properties of both pellicular and conventional ion exchangers having the same chemical composition are essentially the same, the literature of ion exchange chromatography can serve as a guide to the selection of a suitable column and eluent system for a given separation problem as well as to the interpretation of the obtained retention pattern. For many specific applications the rapidly growing literature of HPLC can be consulted as pellicular ion exchangers have been widely used in recent years. In the biochemical and biomedical applications of HPLC pellicular anion exchange resins have been particularly useful since suitable anion exchangers for the separation of nucleic acid constituents have not been available in microparticulate form. A wealth of information on this subject can be found in the recent book of Brown.[62] The practical aspects of high performance ion exchange chromatography are discussed and the specifications of a number of commercially available pellicular ion exchangers are listed in a very recent book by Snyder and Kirkland.[63]

The commercially available pellicular ion exchangers offer the broadest line of high performance column packings for ion exchange chromatography and are used for the analysis of a wide variety of samples. The chemical composition and pore structure of many pellicular ion exchangers are significantly different from those of the commonly used conventional resins. Thus superior selectivity can often be obtained for certain solute mixtures with a properly selected pellicular ion exchanger.

Generally the major difference between the operating conditions with conventional and pellicular ion exchanger columns is the eluent composition. Because of the relatively small ratio of the stationary phase to mobile phase volume, weaker eluents suffice with the latter columns. Both salt elution and elution with acids or bases have been employed, but salt elution is preferred for longer column life. In view of the recently established dynamic role of buffers in facilitating the diffusion of weak acids and bases in heterogeneous media,[64] the use of suitable buffers can result in improved column efficiency.

Gradient elution is preferred in the analysis of samples containing many components because it provides higher peak capacities than isocratic elution. Thus, late peaks can be eluted faster and evaluated easier. The increase of the salt concentration in the eluent during elution appears to be the most convenient method to obtain fast separations, although pH gradient or the combination of concentration gradient and pH gradient have frequently been employed.

PERSPECTIVES

Despite the growing employment of microparticulate packings, pellicular sorbents are likely to continue their important role in HPLC. A variety of potentially efficient stationary phases such as inorganic gels, carbohydrate type sorbents, which in bulk form do not give stable column packing for high pressure chromatography, may be utilized in pellicular configuration. On the other hand the separation of large solute molecules requires stationary phases having a wide pore structure and the pellicular structure can confer the mechanical stability necessary for making high performance columns. Thus, the possible development of affinity chromatography for the analysis of minute samples could also profit from this structure. A variety of enzymes have already been prepared in pellicular form[65,66] and such "enzyme resins" are particularly useful in packed bed operations.

Although the particle size of the presently available pellicular ion exchangers is much greater than that of the microparticulate column packings, there is no reason why micropellicular ion exchangers having 10 μm particle diameter cannot be prepared. As long as the rate of solute exchange underlying the chromatographic process is controlled by diffusion, $h_S/h_M > 1$, the theoretical advantages of the pellicular structure can be exploited irrespectively of the particle size. The performance of micropellicular columns would be superior to that of conventional microparticulate columns and their loading capacity would be significantly higher than that of the present pellicular ion exchanger columns. The reduction of the intraparticular diffusion distances, however, may not result in a further increase in the column efficiency if the solute exchange is not diffusion but kinetically controlled. Thus, the slowness of the desorption rate may set a lower practical limit to the particle diameter or shell thickness.

Whereas ion exchange phenomena with small inorganic ions are fairly well understood, the interaction of relatively large and electrically charged organic molecules with ion exchangers requires further investigation. At present very little data are available on the sorption kinetics and transport of complex electrolytes in ion exchangers. After 25 years of successful practical application in the laboratory, ion exchange chromatography of complex organic molecules is still an arcane art rather than a scientifically established technique. In the past, however, many kinetic effects pertaining to the interaction of the solutes with the ion exchanger and with each other were masked by the slow diffusion rates which controlled the dynamics of the process. It is expected that the attenuation of diffusional effects in small ion exchanger particles will make it possible to observe such kinetic processes and thus to gain a better understanding of the physicochemical phenomena involved. On the other hand, the recognition of such phenomena will aid the design of stationary phases with superior efficiency for the separation of organic molecules. It can be inferred from the presently available data that rigid porous ion exchangers with a uniform pore structure and spatial distribution of the functional groups would give the best results. In addition the arrangement of the relative positions of the various groups on the surface is expected to play a significant role, particularly in ion exchange chromatography of large multifunctional molecules.

The concept of pellicular ion exchangers is not restricted to spherical particles. Small bore tubes with an ion exchanger layer on the inner wall could also find application in chromatography and related techniques. Although capillary tubes with a thin ion exchanger lining gave disappointing results in the past because of the excessive band spreading due to mobile phase effects in single-phase laminar flow, recent measurements with slug flow in similar enzyme coated tubes[67] suggest that small bore tubes with an ion exchanger wall could yield a viable chromatographic system when operated with slug flow which augments radial mixing at low flow velocities.

ACKNOWLEDGMENTS

The author appreciates the information supplied by W. P. Taggart of H. Reeve Angel and Co., Inc. and J. Little of Waters Associates Inc. This work was supported by grants GM 21084-01 and RR-0356 from the National Institutes of Health and the U.S. Public Health Service.

REFERENCES

1. Helfferich, F. In *Advances in Chromatography*, Vol. I, J. C. Giddings and R. A. Keller, eds., (New York: Dekker, 1965), pp. 3-60.
2. Samuelson, O. *Z. Analyt. Chem., 116*, 328 (1939).
3. Johnson, W. C., L. L. Quill, and F. Daniels. *Chem. Eng. News, 25*, 2495 (1947).
4. Cohn, W. E. and H. W. Kohn. *J. Am. Chem. Soc., 70*, 1986 (1948).
5. Moore, S., D. H. Spackman, and W. H. Stein. *Anal. Chem., 30*, 1185 (1958).
6. Anderson, N. G., J. G. Green, M. L. Barber, and F. C. Ladd, Sr. *Anal. Biochem., 6*, 1953 (1963).
7. Scott, C. D., J. E. Attril, and N. G. Anderson. *Proc. Soc. Exptl. Biol. Med., 125*, 181 (1967).
8. Horvath, C. In *Methods of Biochemical Analysis*, Vol. 21, D. Glick, ed. (New York: Wiley, 1973), pp. 79-154.
9. Horvath, C. In *Ion Exchange and Solvent Extraction*, Vol. 5, J. A. Marinsky and Y. Marcus, eds. (New York: Dekker, 1973), pp. 207-260.
10. Rothbart, H. L. In *An Introduction to Separation Science*, B. L. Karger, L. R. Snyder, and C. Horvath, eds. (New York: Wiley, 1973), pp. 337-373.
11. Kunin, R. *Ion Exchange Resins*, 2nd ed. (New York: Wiley, 1958).
12. Wheaton, R. M. and M. J. Hatch. In *Ion Exchange*, Vol. 2, J. A. Marinsky, ed. (New York: Dekker, 1969), pp. 191-234.
13. Patterson, J. A. In *Biochemical Aspects of Reactions on Solid Supports*, G. R. Stark, ed. (New York: Academic Press, 1971), pp. 189-213.
14. Amphlett, C. B. *Inorganic Ion Exchangers*. (Amsterdam: Elsevier, 1964).
15. Helfferich, F. *Ion Exchange*. (New York: McGraw-Hill, 1962).
16. Inczedy, J. *Analytical Applications of Ion Exchange*. (New York: Pergamon Press, 1966).
17. Samuelson, O. *Ion Exchange Separations in Analytical Chemistry*. (New York: Wiley, 1963).
18. Marinsky, J. A. and Y. Marcus, eds. *Ion Exchange and Solvent Extraction, A Series of Advances*, Vol. 3-5 (New York: Dekker, 1973).
19. Marinsky, J. A., ed. *Ion Exchange, A Series of Advances*, Vol. 1 and 2 (New York: Dekker, 1966, 1969).
20. Walton, H. F. In *Chromatography*, E. Heftman, ed. (New York: Reinhold, 1967), pp. 287-342.
21. Hamilton, P. B. In *Advances in Chromatography*, Vol. II J. C. Giddings and R. A. Keller, eds. (New York: Dekker, 1966), pp. 3-60.

22. Cesarano, C. and G. Pagetti. *Comit. Natl. Energia Nucl. RT/CH 63*, 4 (1963).
23. Hamilton, P. B., D. C. Bogue, and R. A. Anderson. *Anal. Chem.*, *32*, 1783 (1960).
24. Scott, C. D. *Clin. Chem.*, *14*, 521 (1968).
25. Burtis, C. A., N. M. Munk, and F. R. MacDonald. *Clin. Chem.*, *16*, 667 (1970)
26. Pepper, K. W., D. Reichenberg and D. K. Hale. *J. Chem. Soc.*,3129 (1952).
27. Weiss, D. E. *Australian J. Appl. Sci.*, *4*, 510 (1953).
28. Parrish, J. R. *Nature*, *207*, 402 (1965).
29. Skafi, M. and K. Lieser. *Z. Anal. Chem.*, *249*, 182 (1970).
30. Skafi, M. and K. Lieser. *Z. Anal. Chem.*, *251*, 177 (1970).
31. Fricke, G. H., D. Rosenthal and G. A. Welford. *Anal. Chem.*, *43*, 648 (1971).
32. Boardman, N. K. *Biochim. Biophys. Acta*, *18*, 290 (1955).
33. Boardman, N. K. *J. Chromatog.*, *2*, 388 (1959).
34. Boardman, N. K. *J. Chromatog.*, *2*, 398 (1959).
35. Feitelson, J. and S. M. Partridge. *Biochem. J.*, *64*, 607 (1956).
36. Horvath, C. and S. R. Lipsky. *Nature*, *211*, 748 (1966).
37. Weiss, J. F. and A. D. Kelmers. *Biochemistry*, *6*, 2507 (1967).
38. Flood, H. *Z. Analyt. Chem.*, *120*, 237 (1940).
39. Randerath, K. *Thin Layer Chromatography*, 2nd ed. (New York: Academic Press, 1966), pp. 229-243.
40. Cassidy, P. E. and B. J. Yager. *J. Macrom. Sci. Revs. Polym. Technol.*, *D 1*, 1 (1971).
41. Aue, W. A. and C. R. Hastings. *J. Chromatog.*, *42*, 319 (1969).
42. Kirkland, J. J. and J. J. De Stefano. *J. Chromatog. Sci.*, *8*, 309 (1970).
43. Majors, R. E. *J. Chromatog. Sci.*, *10*, (1973).
44. Horvath, C. G., B. A. Preiss and S. R. Lipsky. *Anal. Chem.*, *39*, 1422 (1967).
45. Horvath, C. and S. R. Lipsky. *Anal. Chem.*, *41*, 1227 (1969).
46. Kirkland, J. J. *J. Chromatog. Sci.*, *7*, 361 (1969).
47. Kirkland, J. J. *J. Chromatog. Sci.*, *8*, 72 (1970).
48. Kunin, R. *Ion Exchange Resins*, 2nd ed. (New York: Wiley, 1958), p. 100.
49. Kirkland, J. J. U.S. Patent No. 3,488,922 (1970).
50. Kirkland, J. J. *J. Chromatog. Sci.*, *7*, 7 (1969).
51. Kirkland, J. J. U.S. Patent No. 3,577,226 (1971).
52. Henry, R. A., J. A. Schmit and R. C. Williams. *J. Chromatog. Sci.*, *11*, 358 (1973).
53. Heywood, H. *Powder Metallurgy*, *7*, 1 (1961).
54. Fairs, G. L. *J. Roy. Microsc. Soc.*, *71*, 209 (1951).

55. Burger, C. L. *J. Chromatog.*, *26*, 334 (1967).
56. Karger, B. L., L. R. Snyder and C. Horvath. *An Introduction to Separation Science.* (New York: Wiley, 1973), pp. 543-545.
57. Pesek, J. J. and J. H. Frost. *Anal. Chem.*, *45*, 1762 (1973).
58. Giddings, J. C. *Dynamics of Chromatography, Part I, Principles and Theory.* (New York: Dekker, 1965) pp. 143-144.
59. Horvath, C. and S. R. Lipsky. *J. Chromatog. Sci.*, *7*, 109 (1969).
60. Done, J. N., G. J. Kennedy and J. H. Knox. In *Gas Chromatography 1972.* S. G. Perry and E. R. Adlard, eds. (Barking, Essex, England: Applied Science Publishers, 1973), pp. 145-155.
61. Scott, C. D. and N. E. Lee. *J. Chromatog.*, *42*, 263 (1969).
62. Brown, P. R. *High Pressure Liquid Chromatography: Biochemical and Biomedical Applications* (New York: Academic Press, 1973).
63. Snyder, L. R. and J. J. Kirkland. *Introduction to Modern Liquid Chromatography.* (New York: Wiley, Interscience, 1974), pp. 283-328.
64. Engasser, J.-M. and C. Horvath. *Biochim. Biophys. Acta*, in press.
65. Horvath, C. and J.-M. Engasser. *Ind. Eng. Chem. Fundam.*, *12*, 229 (1973).
66. Horvath, C. *Biochim. Biophys. Acta*, in press.
67. Horvath, C. , B.A. Solomon and J.-M. Engasser. *Ind. Eng. Chem. Fundam.*, *12*, 431 (1973).
68. Rieman, W. and H. F. Walton. *Ion Exchange in Analytical Chemistry* (New York: Pergamon Press, 1970).

CHAPTER 6

SUPPORT MATERIALS FOR IMMOBILIZED ENZYMES AND AFFINITY CHROMATOGRAPHY

Garfield P. Royer and William E. Meyers

INTRODUCTION

The recent interest of biochemists in affinity chromatography and immobilized enzymes has been intense. We believe that this interest will go beyond biochemistry when isolation and enzymatic transformation of nonbiological compounds is developed. Analysis and isolation of nonbiological compounds with immobilized antibodies is already feasible. Enzymatic catalysis of reactions of synthetic compounds which resemble biological substrates is certainly possible in principle. With these thoughts in mind the purpose of this chapter is to describe affinity adsorbents and enzyme supports for a general audience, to discuss the "state-of-the-art," and to describe some recent work in our laboratory on support materials.

The principle of affinity chromatography is the insolubilization of a molecule for which the desired compound has a biochemically specific attraction. Use of antigens for the purification of antibodies was performed as early as 1951 by Campbell, Leuscher and Lerman.[1] The process of immunoadsorption later evolved to affinity chromatography when antibodies with varying affinities for the antigen were fractionated.[2] Lerman[3] demonstrated that preparations of the enzyme tyrosinase could be enriched by a substrate-like material, p-azophenol-cellulose. The first example of affinity chromatography as we usually think of it today was the work of Arsenis and McCormich.[4] These investigators demonstrated dramatic purification of the enzyme flavokinase with columns of carboxymethyl cellulose substituted with isoalloxazine derivatives, I and II

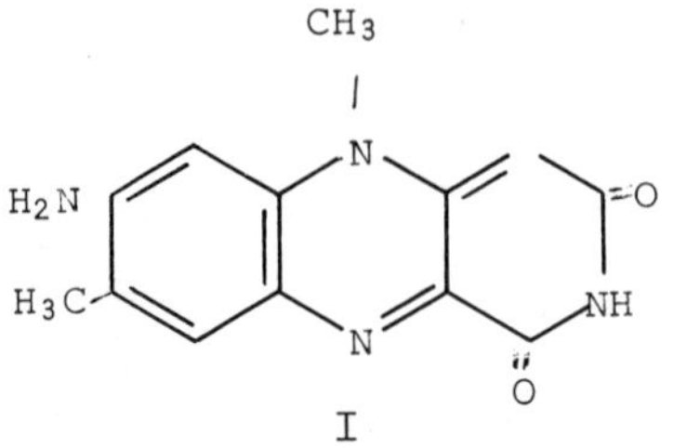

I

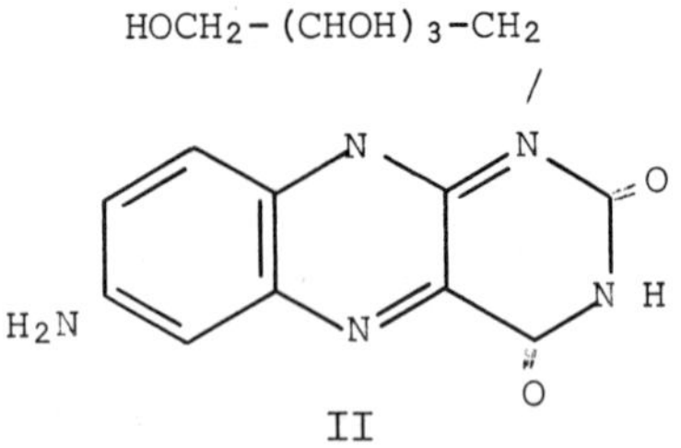

II

These compounds resemble the flavin coenzyme used by this enzyme. Affinity columns which contain coenzymes and inhibitors or their derivatives are the most popular at present. The approach of Arsenis and McCormich was made general in application by the introduction of the anchoring arm by Cuatracasas, *et al.*[5] Specific examples may be found in a number of recent review articles.[6-8]

Enzyme immobilization may be defined as covalent attachment to, adsorption to, or entrapment within an insoluble support. Intermolecular cross-linking of enzyme molecules to form an insoluble aggregate is also a method of immobilization. Only covalent attachment of enzymes to insoluble supports will be considered here.

Why immobilize an enzyme? Most enzymes occur *in vivo* attached to membranes or fixed in multi-enzyme complexes. It is, therefore, important to consider the condition of freely diffusing substrate and stationary enzyme or enzymes. It has been shown in the laboratories of Mosbach[9] and Katchalski[10] that two-enzyme systems supported on a solid or in a membrane behave differently than the systems in which the enzymes are free in solution. Consider the following scheme:

$$A \xrightarrow[\text{enzyme 1}]{} B \xrightarrow[\text{enzyme 2}]{} C$$

The product of the first reaction catalyzed by enzyme 1 serves as the substrate for enzyme 2 which catalyzes the formation of C. Immobilization of enzymes 1 and 2 on the same support considerably reduces the lag time in the appearance of C. When multiplied over a metabolic sequence involving many reactions, this contribution is no doubt very important in the rapid responses and control in metabolic pathways.[11]

Interest in applications of enzymes is based on the fact that they are the most specific and effective catalysts known. Immobilization permits rapid separation of enzyme from reaction mixture and reuse. The stability of immobilized enzymes is often superior to soluble enzymes. A dramatic example of the application of immobilized enzymes in steroid transformations was cited by Mosbach[12] (Figure 6.1). Specific introduction of functional groups into complex steroids with good yields and limited byproducts is no easy task. Much tedious synthetic work may be replaced by immobilized enzyme systems. The

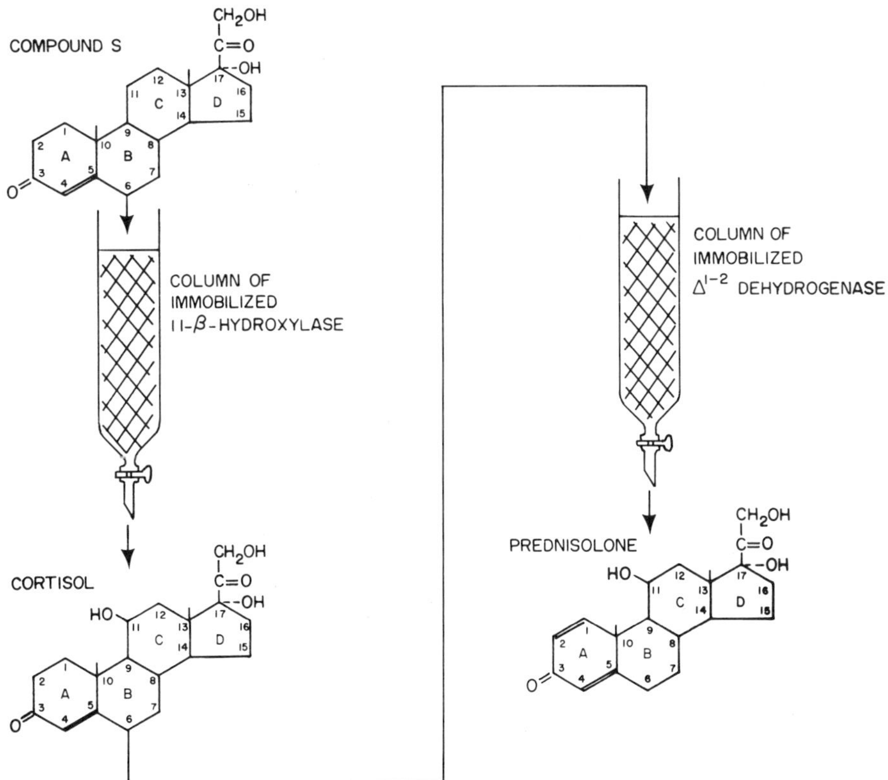

Figure 6.1. Conversion of compound S to prednisolone using immobilized 11-β-hydroxylase and Δ^{1-2} dehydrogenase. Enzyme immobilization allows for rapid separation of reaction mixture from catalyst and reuse of catalyst.

inexpensive starting material, compound S (Figure 6.1), is passed over a column of immobilized 17β-hydrolylase. The product, cortisone, may then be converted to a more effective drug by introducing unsaturation at the 1-2 position. This latter step is accomplished by treatment with Δ^{1-2} dehydrogenase.

Of obvious importance in preparation of affinity columns and immobilized enzymes is the support to which the ligand or enzyme is attached. A support suitable for one technique is often suitable for the other. Important points in the development of the affinity chromatography supports are:

1. they may be of use directly for enzyme immobilization (*i.e.*, without further derivatization)
2. any improvement in affinity chromatography would lower enzyme costs which would in turn expand the application of immobilized enzymes.

There are a great many supports and coupling methods available for affinity ligands and enzymes.[6-8,13-] Two of the more popular supports are agarose and porous glass. Important advantages of these supports are macroporosity and hydrophilicity. The glass has excellent dimensional stability and is resistant to microbial attack. There are a great many ways to activate agarose and porous glass. Some common coupling routes for agarose are given in Figure 6.2 Porous glass is activated as shown in Figure 6.3. In addition enzymes have been linked successfully to alkylamine glass through the bifunctional reagent glutaraldehyde.[16] The coupling routes described here and their variations are well established and applicable to a wide variety of ligands, enzymes, and supports. It seems reasonable, therefore, to progress to other support requirements.

SUPPORT CRITERIA

The characteristics of an ideal support material would include:

1. chemical functionality — or route thereto — sufficiently general to permit reaction with a variety of groups on proteins and ligands for affinity chromatography
2. low cost

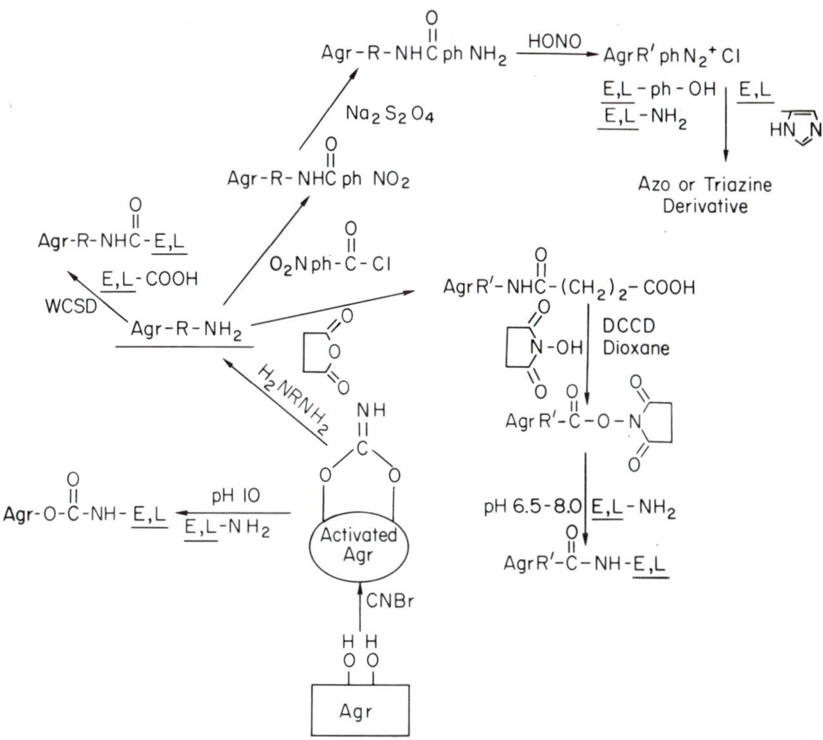

Figure 6.2. *Routes for the covalent attachment of ligands and enzymes to agarose.*

Abbreviations: Agr, agarose; E,L, enzyme or ligand; CNBr, cyanogen bromide; WSCD, water-soluble carbodiimide; DCCD, dicyclohexyl carbodiimide.

Coupling of proteins to glass involves an initial silanization step in which the glass is reacted with a siloxane.

$$H_2N{-}R{-}\underset{\underset{R}{|}\atop O}{\overset{\overset{R'}{|}}{\overset{O}{|}}{Si}}{-}OR' + {}^{-}O{-}Si{-}O{-}(Glass) \longrightarrow$$

$$H_2N{-}R{-}\underset{\underset{R}{|}\atop O}{\overset{\overset{R}{|}}{\overset{O}{|}}{Si}}{-}O{-}Si{-}O(Glass)$$

where R and R' are alkyl groups. The amino group may subsequently be reacted directly with a protein by means of a dehydrating agent such as carbodiimide.

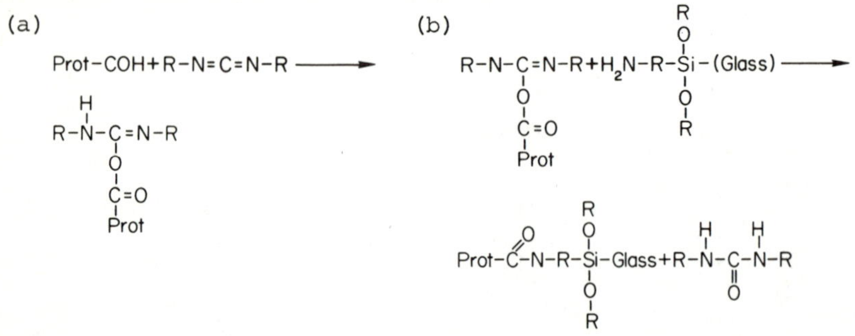

The glass-amine derivative may also be reacted to form intermediates which react with the other groups of the proteins. Among the possible approaches, the diazonium salt (a) and the isocyanate methods (b) are given below:

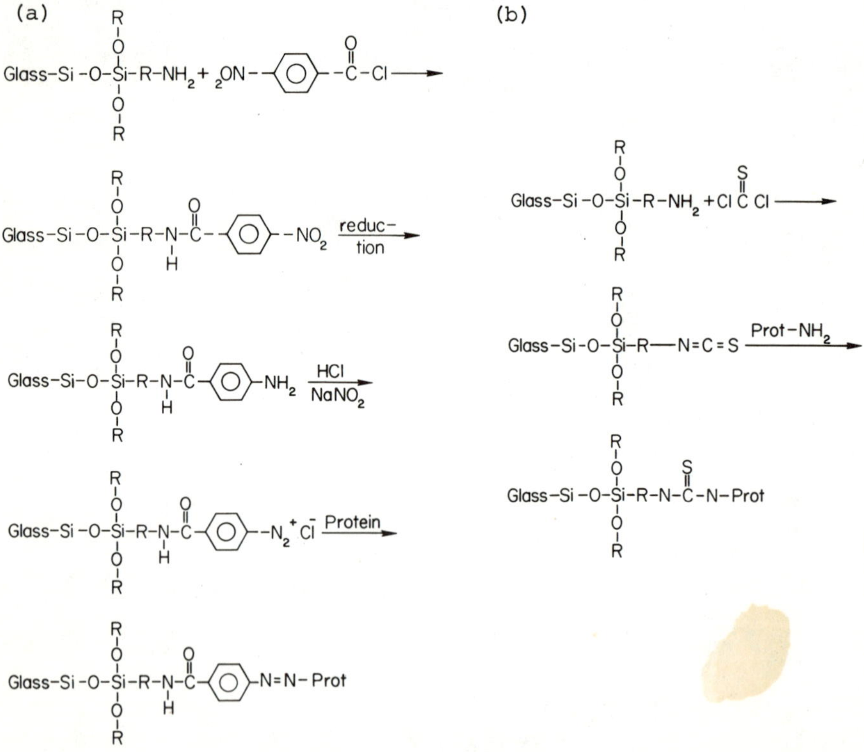

Figure 6.3. Silanization and derivatization of porous glass.

3. resistance to microbial attack
4. dimensional stability — the retention of physical shape following changes in pressure, temperature, solvent composition, etc.
5. durability — the support material should not erode or react chemically with the solvent
6. hydrophilic — many proteins denature at hydrocarbon-water interfaces (*e.g.*, reference 9)
7. regenerable — the feature of carrier reuse could be advantageous. However, if the support is of low initial cost, this may not be critical
8. high capacity for enzyme or ligand
9. accessibility to solvent
10. low level or nonspecific binding.

Porous glass meets most of these requirements. The problems which remain for this support are nonspecific adsorbtion, erosion, and high cost. The first may be solved by various coating techniques. The stability of glass is notably improved by zirconium oxide coating which, unfortunately, also raises the cost. Glass may be regenerated by firing and resilanization. This feature along with the other advantages makes porous glass a very attractive support for many applications.

Agarose has the major advantages of high capacity, virtually no nonspecific adsorption, and hydrophilicity. The major difficulties with agarose are compressibility, high cost, and susceptibility to microbial attack. These problems are severe for many applications of sepharose-bound enzymes. In affinity chromatography, however, the cost is normally justified, antimicrobial agents such as sodium azide may be added, and for many applications batch type reactors would seem feasible.

Agarose and glass are popular but certainly not the only supports available. In the next section we describe and classify other supports.

CLASSIFICATION OF SUPPORTS

The materials used for enzyme and affinity ligands are many and varied. Some are listed below:

Organic Supports

- Carbohydrates
- Vinyl polymers
- Amino acid polymers and derivatives
- Nylon and polystyrene tubes
- Phenol-formaldehyde resins

Inorganic Supports

- Porous glass
- Nickel screen
- Alumina-silicates

Inorganic-Organic Supports

Separate sections of this chapter are devoted to phenol-formaldehyde resins and inorganic-organic supports. The other carriers listed above are reviewed elsewhere.[15]

PHENOL-FORMALDEHYDE RESINS

We have recently investigated a macroporous phenol-formaldehyde resin (Duolite A-7) as a support for enzymes.[17] Attractive features of this material include: macroporous structure, low cost, dimensional stability, chemical stability, and resistance to microbial attack. The resin, which contains amino groups, may be derivatized in a variety of ways (Table 6.1). The most active derivative is the azo-linked trypsin. The resin is alkylated with p-nitro benzyl chloride and reduced with dithionite. Immediately prior to coupling, the material is diazotized and washed with sulfamic acid. In 70 minutes this derivative binds 13 mg trypsin per g as judged by activity depletion from the supernatant. The coupling reaction was carried out as previously described.[18] Studies on dinitrophenyl Duolite A-7 as an immuno-adsorbent are now in progress in our laboratory.

The results in Table 6.1 and preliminary studies indicate that the phenol-formaldehyde resin should be of great value as a support for enzymes and affinity ligands. The material is considerably less expensive than porous glass or agarose but has many advantages in common with these supports. Experiments to evaluate the general applicability of phenol-formaldehyde resins are in progress.

Table 6.1

Activity of Trypsin Bound to Phenol-Formaldehyde Resin and Porous Glass

Support	*Derivative*	*Coupling Reaction*	*Specific Activity*[a]
Duolite A-7	unmodified	glutaraldehyde	0.9
	$-NH-(CH_2CH_2)_n$ COOH	CMC[b]	11.4
	$-NH-COCH_2-CH_2-COOH$	CMC[b]	21.7
	$-NH-COCH_2-CH_2-COOH$	N-hydroxysuccinimide ester	10.3(23)[c]
	-pH-NH_2	Diazonium Salt	109
Porous Glass	$Si(OR)_2R'NH\overset{O}{\overset{\parallel}{C}}-pHNH_2$	Diazonium Salt	100(110)

[a]μmoles/min/g carrier, substrate is 1 mM benzoylarginine ethyl ester, at pH 8.0, 25 mM $CaCl_2$ and 25°.

[b]1 cyclohexyl-3-(2-morpholinoethyl)-carbodiimide metho-p-toluene sulfonate.

[c]Values in parentheses were determined at pH 9.5.

INORGANIC-ORGANIC COMPOSITES

There are available many organic carriers for binding affinity ligands and enzymes.[15] For applications in packed-bed and batch reactors a rigid support is usually desirable. By attaching organic polymers to a rigid inorganic substance, one could presumably combine advantages of both types of carriers. Our approach was to attach various polymers to porous and nonporous silicas. In the case of the porous silica the principal objective was to amplify binding capacity of the support by attaching a polymer with many potential binding groups. The nonporous silicas have the advantages of very low cost and, at small particle sizes, very large surface area.

The structure of polyethyleneimine (PEI) appears in Figure 6.4. The polymer is inexpensive, globular in shape, and readily modifiable with straightforward reactions.[19-21] PEI was bound to the alkylamine derivative of porous glass with glutaraldehyde. The product was acylated with p-nitrobenzoyl

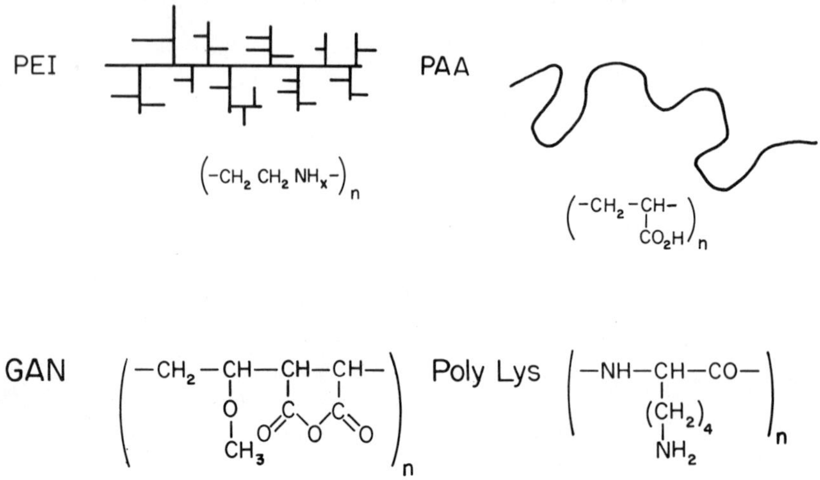

PEI $(-CH_2CH_2NH_x-)_n$

PAA $\left(-CH_2-\underset{CO_2H}{CH}-\right)_n$

GAN (—CH₂—CH—CH—CH—)ₙ with OCH₃ and anhydride ring

Poly Lys $\left(-NH-\underset{\underset{NH_2}{(CH_2)_4}}{CH}-CO-\right)_n$

Figure 6.4. Structures of polymers used for inorganic-organic supports. PEI, polyethyleneimine; PAA, polyacrylic acid; GAN, poly(methyl vinyl ether/maleic anhydride); Poly Lys, polymer of α-L-lysine.

chloride; the nitro group was then reduced with dithionite. The resulting arylamine is diazotized and reacted with trypsin.[18] To eliminate the possibility of intraparticle reaction of a diazonium salt with a neighboring primary amine, we treated the p-nitro derivative with formaldehyde and sodium borohydride.[22] The nitro group was then reduced as before. The results of the activity determinations for trypsin derivatives of these carriers are shown in Table 6.2. Both supports are less effective than the starting material, porous glass. A linear polymer, polyacrylic acid (Figure 6.4), was attached to porous glass with similar results (Table 6.2). The most reasonable explanation for this failure is the reduction of available surface area. The polymers could well block pores leading to the interior of the particle.

An alternative approach is to adsorb PEI onto silica beads and stabilize it with glutaraldehyde (Figure 6.5). This PEI-coated silica is attractive for several reasons: 1) the support is inexpensive and easy to prepare; 2) it may be filtered or centrifuged; 3) it has excellent storage and operational

Table 6.2

Summary of Activities of Trypsin Conjugates with Polymer Coated Silica

Support	*Activity at pH opt. (Units / derivative)*
Porous glass	110
Glass-PEI[a]-Ph-NH_2	69
Glass-PEI(CH_3)-Ph-NH_2	55
Glass-PAA[b]	59
Silica-PEI-GAN[c]	82
Silica-PEI-PAA	300

[a]PEI: polyethyleneimine

[b]PAA: polyacrylic acid

[c]GAN: poly (methyl vinyl ether/maleic anhydride)

stability, and 4) it may be readily derivatized or used directly for coupling of enzyme or ligand.

We reacted poly(methyl vinyl ether/maleic anhydride) (abbr. GAN) with the silica-PEI (Figures 6.4 and 6.6). Enzyme may be coupled through amino groups to the resulting derivative. The trypsin-GAN-PEI-silica derivative was more active than the derivatives previously described (Table 6.2).

Polyacrylic acid (PAA) can be bonded to PEI-silica in the presence of a water-soluble carbodiimide (Figure 6.6). Proteins or ligands can then be coupled to the remaining carboxyl groups with a second carbodiimide reaction. Trypsin immobilized in this manner has more than three times the activity of the trypsin immobilized on porous glass. Similar amounts of protein are bound to the silica-PEI-PAA (19 mg/g) and the porous glass (23 mg/g). The apparent Michaelis constants are 1.04 mM and 1.38 mM, respectively.

The pH dependence of trypsin bound to silica-PEI-PAA is compared to free trypsin in Figure 6.7. The pH optimum of the bound enzyme is two pH units higher than the free enzyme. Similar observations were made by Goldstein, Levin and Katchalski[23,24] for trypsin and chymotrypsin bound to an ethylene-maleic anhydride copolymer. Our finding indicates

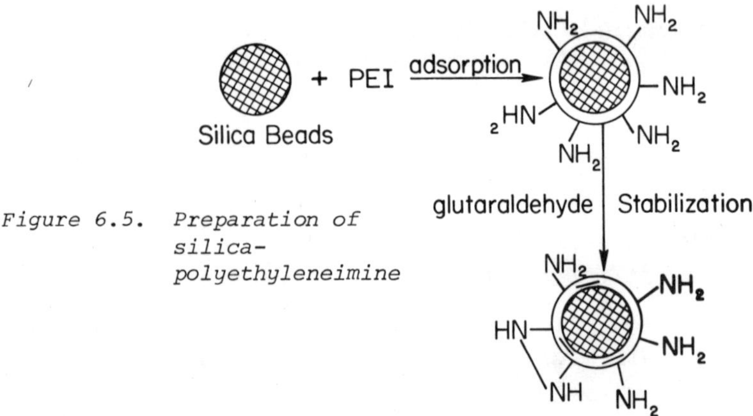

Figure 6.5. Preparation of silica-polyethyleneimine

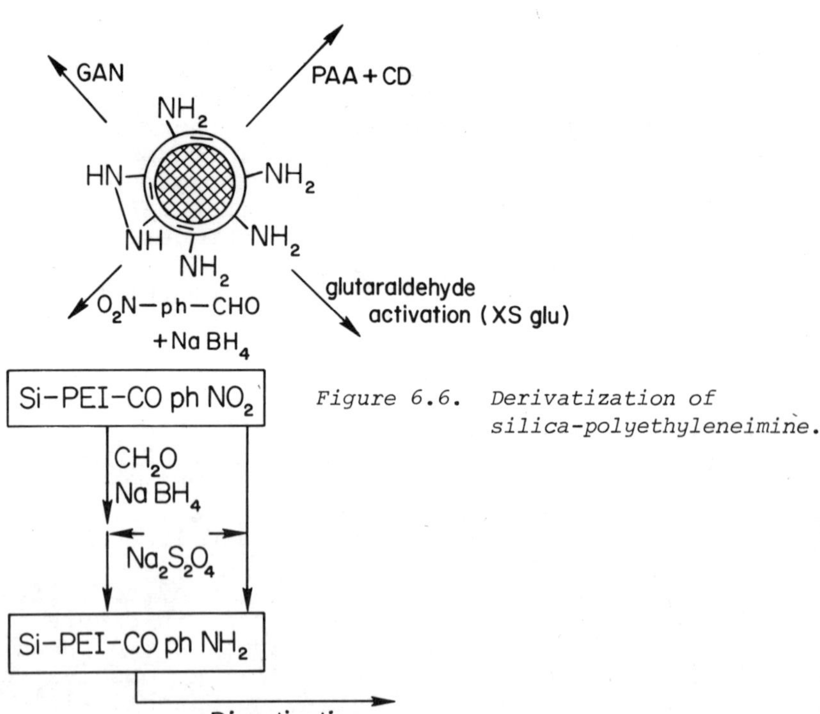

Figure 6.6. Derivatization of silica-polyethyleneimine.

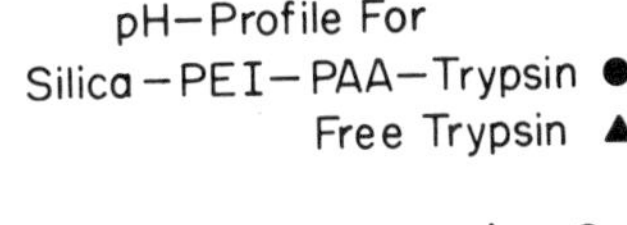

Figure 6.7. *The pH dependence of the hydrolysis of N-α-benzoyl-arginine-ethylester by free trypsin and trypsin bound to silica-polyethylene-imine-polyacrylic acid.*

that the PAA is the predominant polymer in affecting the environment of the enzyme. The picture of the linear PAA projecting as a "whisker" from the bead (Figure 6.6) is therefore reasonable.

POLYMERIC ANCHORING ARMS IN AFFINITY CHROMATOGRAPHY

In the preceding discussion reference has been made to "anchoring arms" in the preparation of some adsorbents for use in affinity chromatography. A more complete discussion of this requirement will be given here with special emphasis on the use of polymers as the linkage between support and ligand. This discussion will help to emphasize the important considerations to be made in choosing the type of linkage to be employed in a given separation.

It might be well to note here that immobilized enzyme systems rarely employ anchoring arms of any length because, as will be seen, they would be of little benefit. Polymeric anchoring arms might be productive, however, if they greatly increase the number of sites for attachment of the enzyme. In some special cases this is of value since multiple site attachment can bind the enzyme into conformations more desirable than the native.[25] Short chains are occasionally incorporated during derivatization, but this is usually limited to the minimum required for the derivatization desired.

In contrast to immobilized enzymes, affinity chromatography often requires anchoring arms of considerable length. Extension of the ligand 10-25Å or more can result in considerable enhancement of column capacities.[26] The advantages derived from the use of anchoring arms may be summarized as follows:

1. Ligand is extended from the vicinity of the matrix and made more accessible.
2. With polymeric anchoring arms the number of functional groups available for ligand attachment may be greatly increased.
3. With polymeric anchoring arms the number of sites of attachment between ligand and support are increased thus reducing the likelihood of disassociation.

Removal of ligand from the vicinity of the matrix may be of two-fold advantage. The first is from purely steric considerations — the "backbone" of the matrix may physically block the approach of a large molecule to the attached ligand not allowing proper orientation for effective bonding. This effect will be especially important when the attached molecule is small and presents only one fixed orientation for bonding. It is this distinction that makes anchoring arms important in affinity chromatography but not in immobilized enzyme systems where large proteins are attached to the matrix.

A second consideration in removing the ligand from the vicinity of the matrix is that the matrix may possess hydrophobic or hydrophilic characteristics which result in an environment unattractive to the molecule to be purified. Effects such as those described above have been observed for many systems, with the resulting effectiveness and capacity directly related to the length of the anchoring linkage.[5,27-31] Cuatracases *et al.*[26] have reported a 5-fold increase in capacity for columns binding staphylococal nuclease by interposing a 13-atom linkage between the ligand, pdTp-aminophenyl, and the support agarose. In our lab we have observed a significant increase in the capacity of columns prepared for the purification of antibodies to dinitrophenyl derivatives by inserting hexamethylenediamine between the agarose support and the nitrophenyl moiety.[31]

$$\text{Agarose}-O-C_6H_3(NO_2)_2 \qquad \text{Agarose}-NH-CH_2-CH_2-CH_2-CH_2-CH_2-CH_2-NH-C_6H_3(NO_2)_2$$

The introduction of long chain bifunctional molecules such as $NH_2(CH_2)_xNH_2$ is an efficient means of preparing many adsorbents for chromatography. Supports with straight chain anchoring arms are now commercially available from Bio-Rad and Pharmacia.

Straight chain anchoring arms like those discussed above take advantage of the removal of the ligand from the vicinity of the matrix but do not exhibit the other possible advantages mentioned for attachment by the anchoring arm, *i.e.*, multiple attachment to the matrix and increased sites for ligand attachment.

To examine these possibilities we have undertaken studies using polylysine as the anchoring molecule for use in the affinity chromatography of antibodies to dinitrophenyl derivatives. Cyanogen bromide activated agarose has been reacted with polylysine to produce a support material with greatly enhanced numbers of primary amine functional groups and multiple sites of attachment between matrix and polymer. The remaining available amines of the polymer are then reacted with 2,4-dinitrofluorobenzene to give an immuno-adsorbent of improved capacity (Figure 6.8). The capacity of this adsorbent per mole of dinitrophenyl moiety is more than twice that of the underivatized agarose adsorbent. This increase is significant but not so great as might have been expected in view of the steric relief allowed by the polymer. The degree of improvement we have observed is limited largely by two factors: crowding of the dinitrophenyl groups on the polylysine and folding of the polylysine into conformations not exposing the ligand moieties. Clustering of the ligand would result in bound antibodies preventing further binding. Both of these problems may be

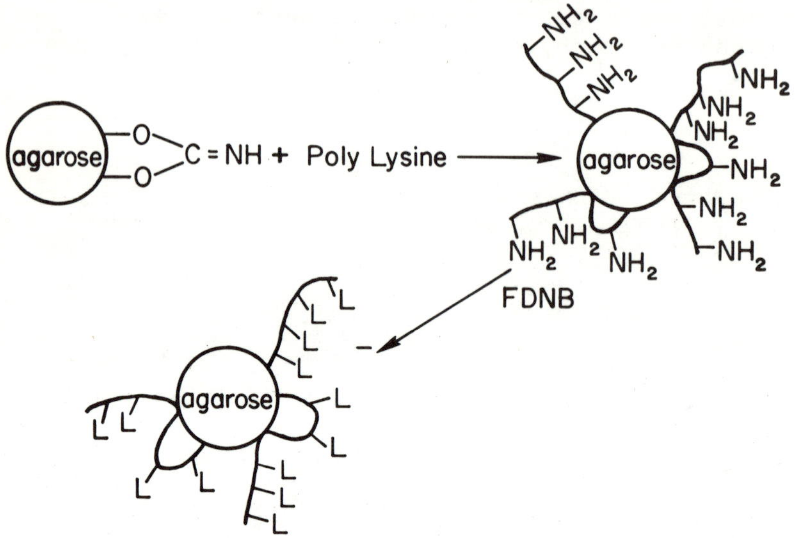

Figure 6.8. *Preparation of an immunoadsorbent for anti-dinitrophenyl immunoglobulins. Polylysine is reacted with cyanogen bromide activated sepharose and the remaining amines are substituted using 2,4-dinitrofluorobenzene.*

explained by the hydrophobic nature of dinitrobenzene. Crowding may be due to a cooperative effect resulting from the effect of one attached dinitrophenyl group attracting a second to that region of the polymer where it may also react. This would result in pockets of dinitrophenylated side chains. The folding problem would arise when two distant pockets of dinitrophenyl groups were drawn to each other by their mutual hydrophobicity. The substituted polymer itself may also exhibit some undesirable hydrophobicity which reduces the degree of interaction with the antibody molecules. These speculations are supported somewhat by the finding that the adsorbent prepared with a hexamethylenediamine anchoring arm is superior in capacity to the polylysine adsorbent.

Cuatracasas and co-workers have used polymeric anchoring arms.[29,30] These workers used a copolymer of lysine and alanine (1:15) with an agarose support in producing an adsorbent for estrogen receptors extracted from calf uterus. This adsorbent was effective even when diluted 25-100 fold with unsubstituted agarose. Even more effective adsorption was obtained when albumin was used as the intermediate between support and ligand.

Specific adsorption is only part of the task in affinity chromatography. It is necessary to elute the bound molecule and recover it in an active form. Figure 6.9 shows an elution pattern for anti-dinitrophenyl immunoglobulins from the sepharose-polylysine-dinitrophenyl adsorbent discussed above. The recovery of antibody is only 50% when elution is performed with 1 molar propionic acid.

This situation is often found where conditions strong enough to disassociate the adsorbed molecule are also irreversibly inactivating.[29,32] This problem may often be avoided by eluting the bound species with solutions of free ligand. For this type of elution it is desirable that the concentrations

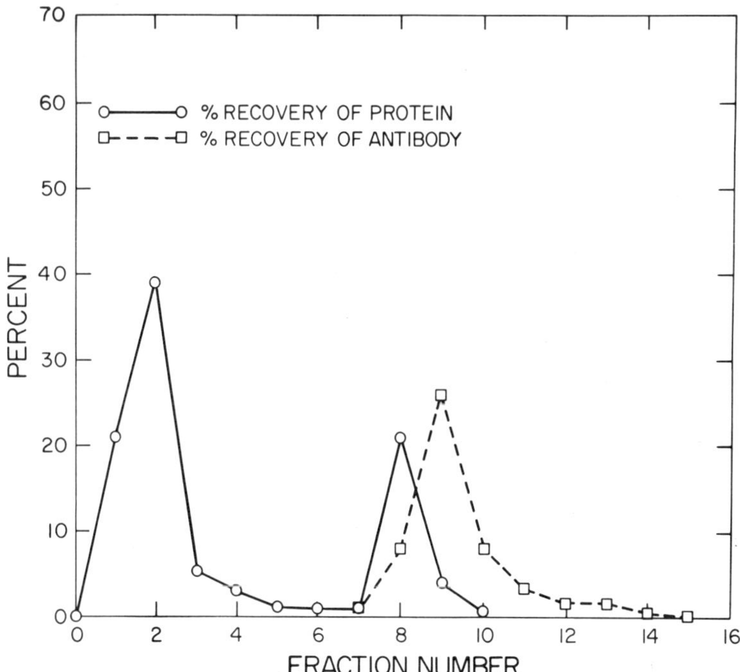

Figure 6.9. Elution of anti-dinitrophenyl immunoglobulins from an adsorbent prepared by dinitrophenylating a polylysine derivative of sepharose. Elution was with 1 M propionic acid. Protein eluates were determined by optical density at 280 mμ, antibody recovery is based on activity measurements.

of free ligand be greater than those of the matrix bound ligand. This procedure has the added advantage of giving an extra degree of specificity to the system; however, it results in somewhat greater elution volumes.

CONCLUSION

For laboratory applications of immobilized enzymes porous glass and agarose are usually satisfactory. Proteolytic enzymes bonded to porous glass may be used for sequencing and total hydrolysis of peptides and proteins.[33-35] Agarose-linked δ-aminolevulinate dehydratase[36] may be used for the synthesis of porphobilinogen with yields far greater than syntheses using conventional nonenzymatic methods or free enzyme. Also, a great variety of enzymes have been purified on bench-top affinity columns based on agarose. Clearly, the demands on a support used in the above applications differ from the demands in a large-scale industrial process.

A low-cost enzyme carrier which would increase substrate binding, decrease product inhibition, shift the apparent pH optimum to the desired value, and discourage microbial growth is often needed. It is our hope that such a material will be found to permit the general application of immobilized enzymes in large-scale processes and to facilitate scale-up of affinity chromatography.

ACKNOWLEDGMENT

NIH Grant GM-19507 supported the work done in the authors' laboratory which is reported here.

REFERENCES

1. Campbell, D. H., E. Leuscher, and L. S. Lerman. *Proc. Nat. Acad. Sci., U.S., 37*, 575 (1951).
2. Cuatracasas, P. *Biochem. Biophys. Res., Commun., 35*, 531 (1969).
3. Lerman, L. S. *Proc. Nat. Acad. Sci., U.S., 39*, 232 (1953).
4. Arsenis, G. and D. B. McCormich. *J. Biol. Chem., 239*, 3094 (1964).

5. Cuatracasas, P., M. Wilchek, and C. B. Anfinsen. *Proc. Nat. Acad. Sci., U.S., 61,* 636 (1968).
6. Cuatracasas, P. in *Biochemical Aspects of Reactions on Solid Supports*, G. R. Stark (Ed.) (New York: Academic Press, 1971), p. 79.
7. Cuatracasas, P. and C. B. Anfinsen. *Annu. Rev. Biochem., 40,* 259 (1971).
8. Cuatracasas, P. and C. B. Anfinsen. *Methods in Enzymol., 22,* 345 (1971).
9. Mosbach, K. and B. Mattiasson. *Acta Chem. Scand., 24,* 2093 (1970).
10. Goldman, R. and E. Katchalski. *J. Theor. Biol., 32,* 243 (1971).
11. Srere, P. A., B. Mattiasson, and K. Mosbach. *Proc. Nat. Acad. Sci., U.S., 70,* 2534 (1973).
12. Mosbach, K. *Scientific American, 224,* 26 (1973).
13. Goldman, R., L. Goldstein, and E. Katchalski. in *Biochemical Aspects of Reactions on Solid Supports*, G. R. Stark (Ed.) (New York: Academic Press, 1971), p. 1.
14. Melrose, G. J. H. *Rev. Pure and Appl. Chem., 21,* 83 (1971).
15. Andrews, J. P., R. Uy, and G. P. Royer. *Enzyme Technol. Digest, 1,* 99 (1973).
16. Weetall, H. H. and N. B. Havewala. in *Enzyme Engineering*, L. B. Wingard, Jr. (Ed.) (New York: John Wiley, 1972), p. 249.
17. Royer, G. P. and G. M. Green. *Polymer Preprints, 15,* 387 (1974).
18. Royer, G. P. and G. M. Green. *Biochem. Biophys. Research Commun., 44,* 426 (1971).
19. Davis, L. E. in *Water-Soluble Resins*, R. L. Davidson and M. Sittig (Ed.) (New York: Reinhold Publishing Corp., 1968), p. 216.
20. Klotz, I. M., G. P. Royer, and A. R. Sloniewsky. *Biochemistry, 8,* 4752 (1969).
21. Klotz, I. M., G. P. Royer, and I. Scarpa. *Proc. Nat. Acad. Sci., 68,* 263 (1971).
22. Means, G. E. and R. E. Feeney. *Biochemistry, 7,* 2192 (1968).
23. Goldstein, L. and E. Katchalski. *Z. Anal. Chem., 243,* 375 (1968).
24. Goldstein, L., Y. Levin and E. Katchalski. *Biochemistry, 3,* 1913 (1964).
25. Royer, G. P. and R. Uy. *J. Biol. Chem., 248,* 2627 (1973).
26. Cuatracasas, P. *J. Biol. Chem., 245,* 3059 (1970).
27. Steers, E., P. Cuatrecasas, and H. Pollard. *J. Biol. Chem., 246,* 196 (1971).
28. Cuatracasas, P., M. Wilchek, and C. B. Anfinsen. *Biochemistry, 8,* 2277 (1969).
29. Sica, V., I. Parikh, E. Nola, G. Puca, and P. Cuatracasas. *J. Biol. Chem., 248,* 6543 (1973).

30. Cuatracasas, P., I. Parikh, and M. Hollenberg. *Biochemistry, 12,* 4253 (1973).
31. Meyers, W. and G. P. Royer. in preparation.
32. Parikh, I., G. Puca, P. Cuatrecasas. *Nature New Biology, 244(132),* 36 (1973).
33. Royer, G. P. and J. P. Andrews. *Polymer Preprints, 13,* 848 (1972).
34. Royer, G. P. and J. P. Andrews. *J. Macromol. Sci., A7(5),* 1167 (1973).
35. Royer, G. P. and J. P. Andrews. *J. Biol. Chem., 248,* 1807 (1973).
36. Gurne, D. and D. Shemin. *Science, 180,* 1188 (1973).

CHAPTER 7

POLYAMIDE CHROMATOGRAPHY

Fredric M. Rabel

INTRODUCTION

The chromatographer is always interested in new media which will help him (or her) in the analysis of an unknown. Of course, silica, the most versatile chromatographic medium, will hardly be replaced with another as broad in scope since such may not exist. However, new media designed to do specific tasks, either more easily or more specifically, occasionally appear in use in the literature. One example is polyamide, more commonly known as nylon.

Polyamide is so unique and holds so much promise as a liquid chromatographic medium, yet it is comparatively new and unknown. In order to acquaint the chromatographer with this medium, a review of pertinent research with polyamide was thought necessary. This chapter will present the state-of-the-art of polyamide chromatography, with many references and separations which will show the novice how to begin this type of chromatography. Many of the references are derived from column and thin-layer chromatographic (TLC) literature, since few liquid chromatographic (LC) references are available. These chromatographic techniques are very closely related, however, so the lessons learned will find easy application.

POLYAMIDE COLUMN CHROMATOGRAPHY

One of the first mentions of the use of polyamide as a chromatographic medium was made in 1955 by Corelli, Liquori and Mele,[1] who applied columns filled with polyamide to the separation of mixtures of xylenols and aliphatic organic acids, using cyclohexane as the mobile phase. It was these researchers who first postulated that polyamide functioned as a partition chromatography medium, the polyamide being the polar stationary phase and the cyclohexane being the nonpolar mobile phase. Also noted was the large capacity of polyamide columns which would allow preparative chromatography.

Grassmann, Hörmann and Hartl,[2] published further results on polyamide column chromatography of phenols using the mobile phases water, water/acetone (4:1) and water/ethanol (1:1). These researchers found that, in general, decreasing the polarity of the mobile phase would increase the retention of the various phenolic compounds. Selectivity was also changed by the varying of mobile phase constituents. They also reported on the large capacity which polyamide exhibited, having separated micromoles of phenols on a small column containing 3 g of polyamide.

A few years passed and a considerable amount of polyamide column chromatography was performed by many workers. In 1963, Endres and Hörmann[3] published an excellent review (with over 61 references) of the research done to that date. The compounds included in this review were phenols, natural products, carboxylic acids, aromatic nitro compounds (including DNP-amino acids and DNP-peptides) and quinones. This review pointed out that the versatility of polyamide was due to its ability to undergo hydrogen bonding to varying degrees. All mobile phases reported were polar; thus some further versatility of this media was yet to be discovered.

POLYAMIDE THIN-LAYER CHROMATOGRAPHY

Interest in polyamide chromatography continued in the 1960's but shifted from column chromatography to thin-layer chromatography. This form of polyamide chromatography was considerably faster, cheaper, and more convenient and, hence, a considerable volume of TLC work was reported. Some of the key TLC research, as it will relate to LC, will be mentioned. A more complete list of polyamide TLC papers is available commercially.[4]

Dual Chromatographic Nature

Early in 1960, Davidek[5] worked with unbound and starch-bound polyamide thin layers to separate fat antioxidants such as gallic esters, BHA, BHT, etc. Continuing the investigation of such compounds, Copius-Peereboom[6] reported on the interesting dual nature of polyamide chromatography. He found that the R_f values of the gallic esters *decrease* with increasing length of carbon chain when using methanol-acetone-water mixtures. This same trend is observed when separating these compounds by reverse-phase paper chromatography using vegetable and paraffin oils as the impregnating agents. Thus it appeared as if the polyamide layer was acting as a nonpolar stationary phase in such mobile phases. However, he found that the R_f values of the gallic esters *increase* with increasing length of carbon chain when using nonaqueous solvent mixtures containing acetic acid [such as light petroleum-benzene-acetic acid (10:10:5)] on polyamide layers. This same trend was observed when chromatographing the same compounds with the same mobile phase on silica gel G thin layers. Likewise, the same trend is also observed when doing paper chromatography of these compounds with chloroform/acetic acid (99:1). From these findings, Copius-Peereboom proposed that the polyamide-acetic acid complex is comparable to the cellulose-water complex and that one could assume a partitioning process between the polyamide-acetic acid complex (polar stationary phase) and the non-polar mobile phase as being responsible for the observed chromatography. Figure 7.1 is a graphic representation of the R_f data presented by this researcher.

Many chromatographers continued working with polyamide thin layers, but the publications of Wang and his co-workers in the 1960's and 1970's best showed the great versatility of this medium. Table 7.1 is a listing of the types of compounds separated by this group on polyamide layers and the solvent system(s) used. Wang also found that many of the compounds separated on polyamide thin layers pointed out the dual nature of polyamide chromatography.

In his research on opium alkaloid separations on polyamide,[7] Wang noted the R_f values of these compounds with cyclohexane/ethyl acetate/n-propanol/dimethylamine (30:2.5:0.9:0.1) as the mobile phase were in the same order as found when these compounds were separated on silica gel. That is, the R_f value increased when the

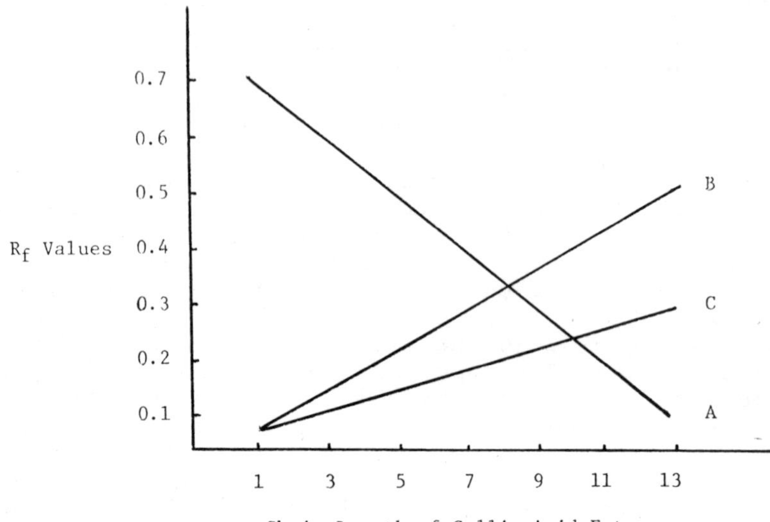

Figure 7.1. *R_f values versus chain length of a series of gallic acid esters[6] in the systems:*
A. polyamide; methanol/acetone/water (60:20:20)
B. polyamide; light petroleum/benzene/acetic acid/dimethylformamide (10:10:5:0.25)
C. silica gel; light petroleum/benzene/acetic acid (10:10:5).

Table 7.1

Polyamide TLC Separations Performed by Wang

Compounds	*Mobile Phase*	*Reference*
Alkaloids	Cyclohexane/EtAc/n-PrOH/ Dimethylamine (30:2.5:0.9:0.1) Water/abs. EtOH/Dimethylamine (88:12:0.1) plus others.	7
Antioxidants	Pet. ether/Benzene/Acetone (8:2:5) Acetone/EtOH/H_2O (4:1:2)	8
Antipyretics	$CHCl_3$/Benzene/90% HFor (5:1:0.1) 2-PrOH/H_2O/90% HFor (1.5:6:0.1) $CHCl_3$/90%/HFor (20:0.1) Cyclohexane/$CHCl_3$/gla. HOAc (4:5:1)	9

Table 7.1, continued

Compounds	*Mobile Phase*	*Reference*
Barbiturates	Hexane/Et_2O/gla. HOAc (1:1:0.01) Hexane/n-BuOH/gla. HOAc (4:1:0.05) plus others.	10
Indoles	$CHCl_3$/EtAc/gla. HOAc (7:2:1) $CHCl_3$/gla. HOAc (9:2) 80% HFor/H_2O (1:2) Et_2O/gla. HOAc (9:2)	11
Lactones	CCl_4/gla. HOAc (9:1) 90% HFor/H_2O (1:2) n-BuOH/90% HFor (9:1)	12
Nitro compounds	CCl_4/gla. HOAc (9:1) 95% EtOH/H_2O (1:1) 90% HFor/H_2O (1:4) n-Hexane/gla. HOAc (4:1) Acetone/H_2O (1:1)	13
Nucleic Acid Bases and Nucleosides	Heptane/n-BuOH/gla. HOAc (4:4:1) CCl_4/HOAc/Acetone (4:1:4) Toluene/Pyridine/Ethylene chlorohydrin/0.8*M* NH_4OH (5:1:5:3:3) 0.5 *M* NaCl	14
Nucleotides	H_2O/gla. HOAc (20:1) 2-PrOH/H_2O/gla. HOAc (2:2:1)	15
Pesticides	Acetone/EtOH/H_2O (2:2:4) EtOH/Ammonia/H_2O (5:2:4)	16
Quinones	H_2O/gla. HOAc (9:1) n-Hexane/Chlorobenzene/gla. HOAc (4:1:0.5) plus others.	17
Sulfonamide	$CHCl_3$/95% EtOH (9:1) EtAc/95% EtOH (8:2) H_2O/95% EtOH (6:4)	18
Thiamines	Acetone Acetone/Et_2O/gla. HOAc (20:20:1) Methyl ethyl ketone $CHCl_3$/EtAc/gla. HOAc (20:20:1)	19
Amino acid derivatives	Benzene/gla. HOAc (8:2) H_2O/gla. HOAc (1:1) plus others.	20-23

hydroxyl group was substituted. This order was reversed, however, when the mobile phase water/absolute ethanol/dimethylamine (88:12:0.1) was used on polyamide.

Wang's work with antioxidants,[8] which includes various gallate esters, also gave results mentioned before by Copius-Peereboom[6] with the solvent systems (1) petroleum ether (30-70°)/benzene/acetone (8:2:5) and (2) acetone/ethyl alcohol/water (4:1:2). In other words, here again the polyamide acts differently, depending upon the polarity of the mobile phase.

When working with lactones[12] Wang compared the R_f values of the compounds coumarin and toddalolactone. In carbon tetrachloride/glacial acetic acid (9:1) coumarin travels faster than does toddalolactone. This is attributed to strong hydrogen bonding of the hydroxyl groups of toddalolactone to the polyamide. In formic acid/water (1:2) the order is reversed and the hydroxyl groups of toddalolactone, being hydrophilic, allow this compound to move faster with the aqueous mobile phase.

Other thin layer chromatographers (see Table 7.2) have reported on the versatility and/or dual nature of polyamide.

Effects of Substitution

Effects of substitution on TLC behavior have also been reported and can be of great use in predicting LC elution order. Graham[24] studied many substituted phenols in polyamide using cyclohexane/glacial acetic acid (93:7) and was able to make the following observations:

1. Intramolecular hydrogen bonding between the phenolic hydrogen atom and an *ortho* nitro group results in an increase in the R_f values relative to isomeric compounds in which the nitro group is found in some other position in the molecule.
2. Polysubstitution with nitro groups results in a reduction of the R_f values.
3. Substitution with halogen atoms also causes a reduction in R_f values.
4. Substitution with alkyl groups generally increases the R_f values.

Table 7.2

Polyamide TLC Separations of Various Compound Types

Compounds	*Mobile Phase*	*Reference*
Artificial Sweeteners (also food preservatives and dyes)	Benzene/EtAc/Formic acid (5:10:2)	25
Catecholamines	Isobutanol/Cyclohexane/gla. HOAc (80:10:7) 2-PrOH/Ammonia (4:1) n-BuOH/H_2O/gla. HOAc (4:1:1)	26
Imidazole	Methyl ethyl ketone	27
Nitrophenols	Cyclohexane/gla. HOAc (93:7)	24
Phenols	Acetone/H_2O (1:1) Benzene/MeOH/gla. HOAc (45:8:4)	28
Plant Auxins	Methyl ethyl ketone/MeOH (99:1) Benzene/Hexane (1:1)	29
Steroids	Hexane/Acetone (4:1) $CHCl_3$/Acetone (4:1)	30
Steroid Glucuronides	MeOH/H_2O/Formic acid (60:35:5) EtFor/MeOH/H_2O/Formic acid (50:20:25:5)	31
Sugars	EtFor/MeOH (8:1) n-BuOH/Acetone/H_2O/gla. HOAc (6:2:1:1)	32,33
Dyes	Pyridine/MeOH/28% Ammonia/H_2O (5:6:1:16) Ammonia/MeOH/H_2O (1:15:84)	34-37

5. When alkyl groups are substituted in the 2- or 6-position, the R_f values increase with an increase in the size of the alkyl group.

Some of the same generalizations were found by this author when doing liquid chromatography on polyamide and these findings will be commented on later.

Effect of Different Polyamide Structures

Graham[24] performed his TLC work on two different commercially available polyamide thin layers and noted that the similarity in R_f values on the two systems [one made of hexamethylenediamino-adipate (Nylon 66), the other being E-polycaprolactam (Nylon 6)] suggests that the carbon skeleton of the polyamide plays little or no part in the chromatographic mechanism.

The differences in commercial polyamide media for thin layers were also described by Grafe and Engelhardt.[33] These chromatographers made layers of six commercial polyamides and separated a mixture of sugars with butanol/acetone/water/glacial acetic acid (60:20:10:10). Although selectivities do differ, it appears that the most useful layers (and hence bonded phases in LC) would be those composed of Nylon 6 (E-polycaprolactam) or Nylon 66 (hexamethylenediamino-adipate) or combinations thereof. Nylon 11 (poly-aminoundecanoic acid) has little selectivity at least for these sugars. This author has also found Polyamide 6 thin layers to be more selective than Polyamide 11 thin layers in almost all separations attempted. It appears that Polyamide 11 might be most useful in separations of higher molecular weight polar species where fewer amide groups per unit area would be desirable so that less adsorption could occur.

Effect of Acid and Base or Temperature

The TLC literature has also pointed out other effects with which one should be familiar in case they also occur in LC. These are the effects of (a) acid or base and (b) temperature on polyamide chromatography. These are best described by the original work:

In their separation of water-soluble acid dyes on polyamide thin layers, Takeshita, Yamashita and Itoh[34] found many dyes tailed when using methanol/water/ammonia mixtures. This problem was overcome by introducing pyridine into the mobile phase. A very elaborate mobile phase study led them to a pyridine/methanol/28% ammonia/water (5:6:1:16) mixture as being best. Many times the addition of an acid or base to a mobile phase will also eliminate a tailing problem in LC.

Marais[32] accomplished fine separation of sugars on polyamide thin layers using ethyl formate/methanol (8:1). He found a 20-30% increase in R_f values when comparing separations accomplished at 23°C with those at 18°C. The effect of temperature on polyamide LC will be discussed later.

POLYAMIDE LIQUID CHROMATOGRAPHY

Having noted the versatility of polyamide as a chromatographic medium, pellicular polyamide liquid chromatographic media were developed by duPont and Northgate Laboratories in 1970. In 1972, H. Reeve Angel & Company acquired the Northgate products and technology to develop the product line to its full potential. This author then undertook a continuing project of finding and studying separations possible on this medium. In 1973, work was published[39] on the Reeve Angel pellicular polyamide product called Pellidon. To date, no work performed on the duPont polyamide has been seen in the literature.

Properties of Pellidon

The Pellidon polyamide is bonded to the prepared surface of glass beads so that a stable, solvent-resistant product results. The surface is irregular, porous, and of uniform thickness as shown by the electron micrograph in Figure 7.2. The Pellidon layer is completely stable to hexane, benzene, diethyl ether, ethyl acetate, acetone, methanol, dimethylsulfoxide, dimethylformamide, 0.1*N* hydrochloric acid and 0.1*N* sodium hydroxide as well as to other mixed mobile phases mentioned subsequently. The product is screened to give an average particle size of 45 microns and has a bulk density of 1.29/cm^3.

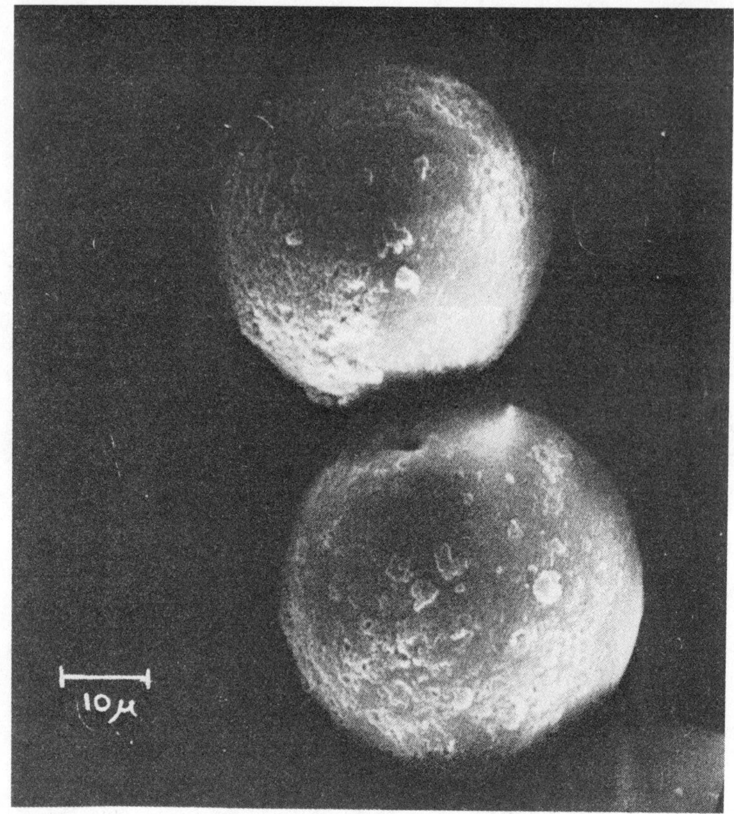

Figure 7.2. Electron photomicrograph of Pellidon (1000x).[39]

Columns of Pellidon are easily prepared by the tap-fill method of Kirkland[40] after the Pellidon is heated for two hours at 50°C. Polyamides tend to adsorb moisture, causing the pellicles to adhere together unless the moisture is removed by slight heating.

Effect of Substitution

The initial study of liquid chromatographic separations which could be performed on Pellidon was done with substituted anilines and phenols. The mobile phases used were hexane with glacial acetic acid and, if a more polar solvent was needed, cyclohexane with acetic acid. The capacity factors derived from this work are given in Table 7.3. A typical chromatogram is shown in Figure 7.3.

Table 7.3

Capacity Factors (k') of Aniline, Phenol and their Derivatives on Pellidon[39]

Solvent System	Aniline	Nitroanilines			Anisidines			Phenetidines		
		o-	m-	p-	o-	m-	p-	o-	m-	p-
n-Hexane/acetic acid (95:5)	0.62	0.85	2.65	14.80	0.40	1.05	5.30	0.30	0.72	3.02
n-Hexane/acetic acid (97.5:2.5)	0.72	1.25	4.35	--	0.25	1.00	7.50	0.10	0.50	3.85
Cyclohexane/acetic acid (90:10)	0.50	0.45	1.05	4.20	0.30	0.30	0.65	0.25	0.25	0.75
Cyclohexane/acetic acid (95:5)	0.68	0.72	2.25	11.65	0.30	0.90	3.68	0.25	0.60	2.15

Solvent System	Toluidines			Phenol	Nitrophenols			Cresols		
	o-	m-	p-		o-	m-	p-	o-	m-	p-
n-Hexane/acetic acid (95:5)	0.25	0.50	1.00	1.12	0.0	--	--	0.50	0.70	0.70
n-Hexane/acetic acid (97.5:2.5)	0.20	0.41	0.85	2.45	0.0	--	--	0.90	1.40	1.40
Cyclohexane/acetic acid (90:10)	0.32	0.32	0.32	0.50	0.0	2.45	4.20	0.0	0.0	0.0
Cyclohexane/acetic acid (95:5)	0.25	0.50	0.95	0.90	0.0	8.90	17.38	0.55	0.75	0.75

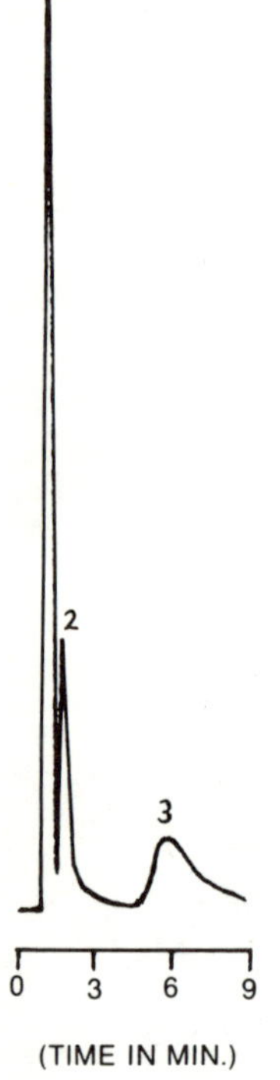

Column, 1-m × 1.8-mm i.d.; sample, 0.5 μl of a mixture of 1 μg each isomer/1 μl in 1-chlorobutane; flow rate, 60.4 ml/hr (140 psi); mobile phase, *n*-hexane/acetic acid (97.5:2.5); detector, UV; temperature, ambient. Peak identity: 1. *o*-phenetidine, 2. *m*-phenetidine, 3. *p*-phenetidine

Figure 7.3. Separation of o-, m-, p-phenetidines on Pellidon.[39]

The order of elution of all substituted aniline and phenol isomers studied was 1) ortho, 2) meta, and 3) para. General patterns of retention on the Pellidon were observed, some similar to those given by Graham[24] and mentioned above. (In comparing the TLC data with the LC data, it must be remembered that a high R_f value for a compound from TLC data will mean a small capacity factor (k') when this same compound is placed on the same system in LC.) These patterns are observed:

1. Substitution of electronegative groups ($-OCH_3$, $-OCH_2CH_3$, $-NO_2$) in the meta or para position of aniline or phenol will increase the k' of the isomer as compared to the unsubstituted parent compound. Such groups can also hydrogen bond to the polyamide, hence k' increases are sometimes quite large.

2. Substitution of electropositive groups (CH_3, $-CH_2CH_3$) in the ortho or meta position of aniline or phenol will decrease the k' of the isomer as compared to the unsubstituted parent compound.
3. Substitution of electronegative groups ($-OCH_3$, $-OCH_2CH_3$, $-NO_2$) in the ortho position of aniline or phenol will decrease the k' of the isomer as compared to the unsubstituted parent compound. This is generally because intramolecular hydrogen bonding can occur in such isomers, hence the substituent groups are less available for hydrogen bonding.
4. Phenol derivatives had larger k' values than their analogous aniline derivatives. This is as expected from infrared data by Flett[41] who determined the energy of hydrogen bonding between phenol and dimethylformamide to be 6.4 kcal/mole, while for benzotoluidine and dimethylformamide it was 3.8 kcal/mole.
5. The order of retention of amines on Pellidon is primary > secondary > tertiary--the latter being unretained (see Table 7.4).

Table 7.4

Capacity Factors (k') of Aryl Amines and Aryl Alcohols[39]
[Pellidon; n-Hexane/gla. HOAc (95:5)]

Compounds	*k'*
Aniline	0.62
N-Methylaniline	0.11
N,N-Dimethylaniline	0.0
Phenol	1.12
Benzyl Alcohol	0.30
α-Methyl benzyl alcohol	0.17
Phenylethyl alcohol	0.0

6. The order of retention of hydroxyl-containing aromatic and nonaromatic systems is phenyl > benzyl > α-methyl benzyl > phenethyl (see Table 7.4). The greater acidity of phenol (as compared to other hydroxyl-containing compounds of nonaromatic nature) is a result of π electron density of the oxygen pulled

> into the π electron system of benzene.[42] The amount of this delocalization, hence, would also control the degree of hydrogen bonding with polyamide. That is, the less the delocalization of the oxygen of the hydroxyl group, the less retention on the Pellidon.

These are reasonable and expected trends if the separations of these compounds on polyamide with such relatively nonpolar mobile phases are viewed as occurring because of hydrogen bonding. This argument is based on considering hydrogen bonding as electrostatic in nature as described by Coulson.[43] Reiser[44] has also discussed ring substitution and its effects on hydrogen bonding.

Recently, a few fluorine-containing compounds were used as samples to see if the fluorine atoms would hydrogen bond to the Pellidon.[45] If such hydrogen bonding did occur, extremely large k' values would be expected; thus the more polar mobile phase of cyclohexane/glacial acetic acid (96:4) was chosen. The k' values found are given in Table 7.5. Apparently, some hydrogen bonding is occurring but not to the degree that one might expect from the compounds chosen. Even inductively, the fluorine atom has not increased the hydrogen bonding of the amino group in p-fluoroaniline as compared to aniline. Perhaps the aromatic molecules chosen are too sterically hindered to fit close enough to the polyamide surface for bonding interactions to occur.

Table 7.5

Capacity Factors (k') of Aryl and Fluorine Containing Aryl Compounds [Pellidon; Cyclohexane/gla. HOAc (96:4)]

Compound	*k'*
Fluorobenzene	0.14
Aniline	0.82
Phenol	1.20
4-Fluoroaniline	0.94
4-Fluoroanisole	0.0
4-Fluorophenol	1.80
Benzoic acid	0.13
4-Fluorobenzoic acid	0.14

Effect of Acetic Acid

Although acetic acid and formic acid were frequently used in mobile phases by other chromatographers, no one apparently has made comment on why the acid was in the mobile phase or what effect varying the concentration of the acid would have on the separation. One of the most interesting findings in the study of Pellidon was the effect of varying the acetic acid concentration in the hexane or cyclohexane. Simply stated, the more acetic acid in the mobile phase, the faster the elution of the compounds which were adsorbed onto Pellidon. Thus, one can obtain more resolution by decreasing the amount of acetic acid in the mobile phase, or speed up a separation by increasing the amount of acetic acid in the mobile phase. This phenomenon is probably caused by competition of the acetic acid molecules and the solute molecules for the polyamide hydrogen bonding sites. Thus, the more acetic acid in the mobile phase, the fewer sites available for solute molecules. The usable acetic acid concentration range was found to be 1 to 20% in hexane or cyclohexane.

Effect of Temperature

Although Pellidon columns are usually run at ambient temperature, the need may arise where a faster separation is desired when using a viscous mobile phase. The effect of temperature on a standard separation was studied from 20°C to 60°C.[45] The data in Table 7.6 shows that the k' values decreased and the N values (number of theoretical plates) increased with increasing temperature.

Table 7.6

Effect of Temperature on k'/N
[Pellidon, cyclohexane/gla. HOAc (9:1)]

	20	*30*	*40*	*50*	*60°C*
o-nitrophenol	0.0/1050	0.0/1110	0.0/1150	0.0/1200	0.0/1270
m-nitrophenol	2.4/460	2.3/530	2.2/600	2.0/630	1.9/680
p-nitrophenol	4.3/420	4.1/500	3.9/520	3.5/550	3.3/600

Since solvents such as chloroform and methylene chloride will tend to swell polyamide,[46] mobile phases containing these solvents should probably be used only at room temperature in case the effects of temperature and pressure might change the surface characteristics. The upper temperature limit was arbitrarily set at 60°C to prevent the surface characteristics (irregular and porous) from changing. Since nylons soften at about 70-90°C under 260 psi,[47] this temperature is within stable use limits.

Typical Separations

The Pellidon study was then continued to more practical applications, many based upon the TLC literature cited above. One such separation was of food preservatives including sorbic acid, methyl-, ethyl-, and i-propyl-p-phydroxyl benzoates with hexane/acetic acid mobile phases.[48] The variation of k' with acetic acid concentrations in the LC separation (see Figure 7.4) is shown in Table 7.7. It was found that methyl benzoate was not retained on Pellidon with hexane/acetic acid mobile phase. Hence the para hydroxyl group of the esters is the active hydrogen bonding site and not the ester carboxyl oxygen. Most interesting is the apparent electronic effect of the various alkyl groups which is changing the retention of the esters. This effect is occurring *through* the aromatic system indicating that delocalization is occurring throughout the molecule. Thus a slight change in inductive effect in one end of the molecule is felt at the other end. Of course, since the para isomer is involved, this effect is more pronounced than it would be in the ortho or meta isomers.

A direct correlation was found for a separation of tryptophan metabolites[48] on Polyamide 6 TLC and LC on Pellidon. The same mobile phase of water/methanol/formic acid (60:35:5) used in TLC produced the chromatogram seen in Figure 7.5 and the data in Table 7.8. Such direct transferring of a TLC separation to LC is the exception, however, and most mobile phases will need some modification to correct for differences in the chromatographic behavior of thin-layer and liquid systems. The open bed and solvent vapor effects in a TLC system are not duplicated in an LC system with the same mobile phase. The LC system is a "closed" system, hence changes caused by the open bed and solvent

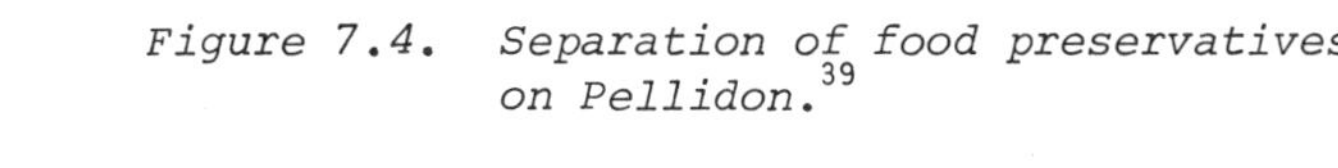

Column, 1-m × 1.8-mm i.d.; sample, 1 μl of a mixture of 0.5 μg each/1 μl in 1-chlorobutane; flow rate, 15 ml/hr (25 psi); mobile phase *n*-hexane/acetic acid (9:1); detector, UV; temperature, ambient: Peak idenity: 1. Sorbic acid, 2. Methyl-*p*-hydroxybenzoate, 3. Ethyl-*p*-hydroxybenzoate, 4. *i*-Propyl-*p*-hydroxybenzoate

Figure 7.4. Separation of food preservatives on Pellidon.[39]

Table 7.7

Capacity Factors (k') of Food Preservatives with Varying Concentrations of Acetic Acid[39]

	n-Hexane/Acetic Acid in Ratio Shown		
Compound/Solvent	(90:10)	(95:5)	(97.5:2.5)
Sorbic acid	0.15	0.15	0.15
Methyl p-hydroxybenzoate	0.80	3.00	6.90
Ethyl p-hydroxybenzoate	1.06	4.15	9.80
i-Propyl p-hydroxybenzoate	1.64	6.88	16.65

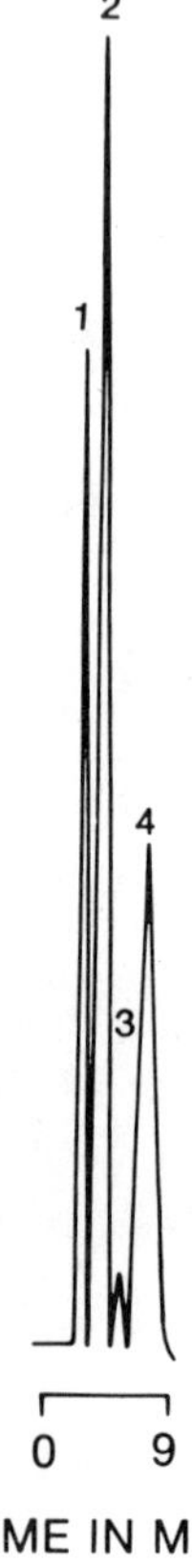

Column, 1-m × 1.8-mm i.d.; sample, 1.5 μl of a mixture of 1 μg each/1 μl in methanol; flow rate, 50 ml/hr (1000 psi); mobile phase, water/methanol/formic acid (60:35:5); detector, UV; temperature, ambient. Peak identity: 1. 3-Hydroxyanthranilic acid, 2. Kynurenic acid, 3. Unknown, 4. Xanthurenic acid

Figure 7.5 *Separation of tryptophan metabolites on Pellidon.*[39]

Vapors in TLC can only be duplicated by suitable solvent modification. From Table 7.8 it can be seen that in going from a TLC to an LC separation some solvent systems remained the same, some were increased in polarity, and some were decreased in polarity. Unfortunately, only experimental work can tell you whether the modification has to be to more or less polarity or to more or less acidity or basicity. Hopefully, as more experience is gained in this area, more guidelines will be drawn.

Table 7.8

Comparison of R_f and k' Values from Polyamide TLC and LC Separations[39]

Compound	*TLC Solvent System,* R_f	*TLC Solvent System,* R_f	*HPLC Mobile Phase,* k'
	Water/methanol/formic acid (60:35:5)[48]		Water/methanol/formic acid (60:35:5)
3-Hydroxyanthranilic acid	0.52	--	0.0
Kynurenic acid	0.28	--	0.50
Xanthurenic acid	0.07	--	1.95
	Water/ethanol (60:40)[18]	Ethyl acetate/ethanol (80:20)[18]	2-Propanol
Sulfamethazine	0.54	0.80	1.20
Sulfaisomidine	0.68	0.65	2.30
Sulfathiazole	0.41	0.38	8.50
	2-Propanol/water/acetic acid (2:2:1)[15]	Acetone/water/acetic acid (2:2:1)[15]	Water/2-propanol/acetic acid (25:5:1)
5'-CMP	0.94	1.00	0.40
5'-AMP	0.56	0.81	0.60
5'-GMP	0.66	0.28	2.50
2',3'-AMP	0.82	0.72	0.60
2',3'-UMP	0.48	0.36	2.30
	Methanol[17]	Methanol/water (9:1)[17]	Water/2-propanol/acetic acid (25:5:1)
p-Benzoquinone	0.67	0.82	0.30
1,4-Naphthoquinone	0.71	0.71	0.50
Anthraquinone	0.57t	0.48	5.10t

Gradient Elution

Gradient elution is a very useful chromatographic technique on any medium. The gradient can be over a small or large polarity range and can be composed of as few as two solvents (or buffers) or as many as twenty (as a practical limit). The literature shows one report of a step gradient elution performed on a polyamide column under gravity flow by Fric and Haspel-Horvatovic.[38] These researchers found polyamide useful in separating plant pigments. Hexane eluted B-carotene, benzene eluted pheophytin and xanthophyll, benzene-chloroform eluted chlorophyll a, and methanol eluted chlorophyll b.

Attempts to do incremental gradient elution on Pellidon have been successful. The gradient technique was derived from the work of Scott and Kucera;[49,50] however, solvent modifications were made so that an ultraviolet (UV) detector (254 A) could be used. Many of Scott and Kucera's solvents, although necessary for smoother gradient, are UV adsorbers; hence they used a moving wire detector.

A Pellidon column (1m x 2.1 mm i.d.) was placed on an LC system with a Milton Roy mini-pump (other low dead volume pumps could be used). The tubing (1/8" o.d. x 2.1 mm i.d. x 1 m) leading from the pump to the septum injector served as the mixing chamber. The solvents were changed every five minutes by stopping the pump, quickly replacing the solvent flask with the next in the series and restarting the pump. The solvent series used for elution and column regeneration are given in Table 7.9. About 15 ml of each regeneration solvent is passed through the column to condition it. A typical separation is shown in Figure 7.6.

Changes in each gradient run are easily made by examining the chromatogram to see what solvents eluted peaks and which ones did not. More or less polar solvent mixtures can be made to expand or compress areas of the chromatogram. It should be remembered that at a flow rate of about 30 ml/hr, it takes approximately six minutes for the solvent introduced into the system to begin to elute from the column. This delay must be considered when examining the chromatogram for possible solvent changes.

This gradient technique is a very useful scouting technique and can easily lead to isocratic or two-solvent gradients which are more desirable for repetitive chromatographic analysis. Pellidon can be a useful alternative to silica gel gradient

Table 7.9

Elution and Reconditioning Solvent Series

Elution Solvents	*Reconditioning Solvents*
Heptane	Ethanol
5% Chloroform in Heptane	Acetone
2% Ethyl acetate in Heptane	Ethyl acetate
1% 2-Propanol in Heptane	1,1,1-Trichloroethane*
5% 2-Propanol in Heptane	Heptane
10% 2-Propanol in Heptane	

*If no trichloroethane is available, use dichloromethane followed by chloroform.

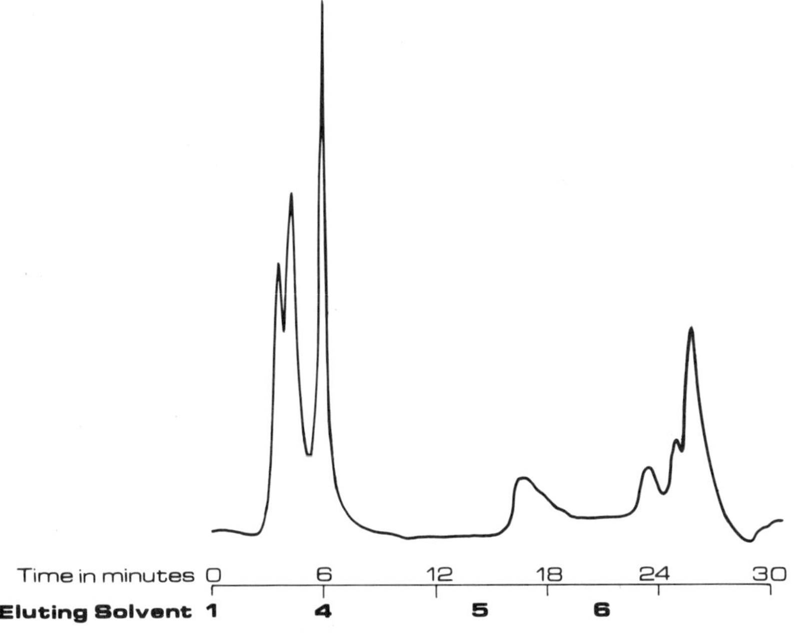

Figure 7.6. *Separation of oak moss resin using gradient elution on Pellidon.*
Column: 1-m x 2.1mm i.d.
Sample: 2 µl of a 2% solution in methanol
Flow Rate: 30.2 ml/hr (100 psi)
Mobile Phases: see Table 8.8
Detector: UV (254)
Temperature: ambient
Peak Identity: none of the peaks have been identified.

elution because of its special selectivity. More details on this technique applied to Pellosil, a pellicular silica gel, are available.[51,54]

Pellidon is a very long-lived packing. Mention has been made that polyamide thin layers can be cleaned and reused, but this is the exception in TLC. Pellicular polyamide, however, is constantly reused as an LC medium. It has been found to be very stable to many solvent changes from polar to nonpolar. Solvent changes can be made rather quickly because only short equilibration times are necessary even when going from one polarity extreme to another. Apparently, few solutes bond themselves irreversibly to polyamide since the regeneration solvents mentioned previously (Table 7.9) always return the Pellidon to its initial chromatographic behavior.

CONCLUSION

The mechanism of the dual nature of polyamide, as derived from TLC data, was discussed in 1968 by Zawta and Hölzel.[52] Essentially, they pointed out that in each polyamide system there is a competition between the solutes and components of the mobile phase for the polyamide sites. If the mobile phase components preferentially adsorb into the polyamide, the solutes can then undergo partitioning between the adsorbed mobile phase component and the mobile phase. This indeed can explain the observed separations discussed previously.

Recently, Soczewinski and Szumilo[53] also began a study of the mechanism of polyamide TLC separations. Their first paper studied the behavior of phenols with cyclohexane modified with polar solvents such as methylene chloride, n-butyl acetate, and C_2-C_5 alcohols. Although the Pellidon separations discussed above made use of acetic acid in nonpolar solvents, the above polar modifiers undoubtedly could be used in place of the acid. An interesting study could be made to describe which modifiers can be used and the differences in the selectivities caused by them.

At this point in time little is known about polyamide separations using water-alcohol or similar polar mobile phases when applied to LC. Although previously described in this chapter by TLC researchers, much work must be accomplished to put the theory into practice. Because of the dual nature, however, an ever-increasing potential exists for polyamide LC separations. Hopefully this review of what has been accomplished and what is known will be beneficial to those in LC research or methods development. With the combination of the old knowledge and that yet to be discovered, the polyamide pellicular packing has the potential to become one of the most interesting of all bonded phase LC packing materials.

REFERENCES

1. Carelli, V., A. M. Liquori, and A. Mele. *Nature, 176,* 70 (1955).
2. Grassmann, W., H. Hörmann, and A. Hartl. *Makromol. Chem., 21,* 37 (1956).
3. Endres, H. and H. Hörmann. *Angew. Chem. Internat. Ed., 2,* 254 (1963).
4. "Bibliography Cheng-Chin Polyamide Layer Sheets" available from Gallard-Schlesinger, Carle Place, N.Y.
5. Davidek, J. and E. Davidkova. *Pharm., 16,* 352 (1961).
6. Copius-Peereboom, J. W. *Nature, 204,* 748 (1964).
7. Huang, J. T., H. C. Hsiu, and K. T. Wang. *J. Chromatog., 29,* 391 (1967).
8. Wang, K. T. and S. S. Chou. *J. Chromatog., 43,* 522 (1969).
9. Hsiu, H. C., T. B. Shih, and K. T. Wang. *J. Chromatog., 41,* 489 (1969).
10. Huang, J. T. and K. T. Wang. *J. Chromatog., 31,* 587 (1967).
11. Wang, K. T., Y. C. Tung, and H. H. Lai. *Nature, 213,* 213 (1967).
12. Wang, K. T. and I. S. Y. Wang. *Nature, 210,* 1039 (1966).
13. Wang, K. T., I. S. Y. Wang, and A. L. Lin. *J. Chinese Chem. Soc., 13,* 77 (1966).
14. Wang, K. T. and I. S. Y. Wang. *Biochim. Biophys. Acta, 142,* 280 (1967).
15. Wang, K. T. and P. H. Wu. *J. Chromatog., 38,* 153 (1968).
16. Wang, K. T. and S. S. Chou. *J. Chromatog., 42,* 416 (1969).
17. Wang, K. T., P. H. Wu, and T. B. Shih. *J. Chromatog., 44,* 635 (1969).
18. Lin, Y. T., K. T. Wang, and T. I. Yang. *J. Chromatog., 20,* 610 (1965).
19. Chuang, H. P., H. C. Chiang, and K. T. Wang. *J. Chromatog., 41,* 487 (1969).
20. Wang, K. T., J. M. K. Huang, and I. S. Y. Wang. *J. Chromatog., 22,* 362 (1966).
21. Wang, K. T., K. Y. Chen, and B. Weinstein. *J. Chromatog., 32,* 592 (1967).

22. Wang, K. T. and I. S. Y. Wang. *J. Chromatog., 27,* 318 (1967).
23. Wang, K. T., I. S. Y. Wang, A. L. Lin, and C. S. Wang. *J. Chromatog., 26,* 323 (1967).
24. Graham, R. J. T. *J. Chromatog., 33,* 118 (1968).
25. Takeshita, R. *J. Chromatog., 66,* 283 (1972).
26. Sapira, J. D. *J. Chromatog., 42,* 134 (1969).
27. Verweij, A. *J. Chromatog., 24,* 473 (1966).
28. Wagner, H., L. Hörhammer, and K. Macek. *J. Chromatog., 31,* 455 (1967).
29. Railton, K. D. *J. Chromatog., 70,* 202 (1972).
30. Wortmann, B., W. Wortmann, and J. C. Touchstone. *J. Chromatog., 70,* 199 (1972).
31. Segura, R., J. Oro, and A. Zlatkis. *J. Chrom. Sci., 8,* 449 (1970).
32. Marais, J. P. *J. Chromatog., 27,* 321 (1967).
33. Grafe, I. and H. Engelhardt. *Chromatographia, 5,* 307 (1972).
34. Takeshita, R., T. Yamashita, and N. Itoh. *J. Chromatog., 73,* 173 (1972).
35. Davidek, J. and E. Davidkova. *J. Chromatog., 26,* 529 (1967).
36. Chiang, H. C. *J. Chromatog., 40,* 189 (1969).
37. Wang, K. T. *Nature, 213,* 212 (1967).
38. Fric, F. and E. Haspel-Horvatovic. *J. Chromatog., 68,* 264 (1972).
39. Rabel, F. M. *Anal. Chem., 45,* 957 (1974).
40. Kirkland, J. J. *J. Chrom. Sci., 10,* 129 (1972).
41. Flett, M. St. C. *J. Soc. Dyers Colour, 68,* 59 (1952).
42. Condon, F. E. and H. Meislich. *Organic Chemistry.* (New York: Holt, Rinehart, and Winston, 1969).
43. Coulson, C. A. *Hydrogen Bonding,* D. Hadzi (Ed.) (New York: Pergamon Press, 1959).
44. Reiser, A. *Hydrogen Bonding,* D. Hadzi (Ed.) (New York: Pergamon Press, 1959).
45. Rabel, F. M. unpublished data.
46. Loev, B. and M. M. Goodman. *Progress in Separation and Purification,* E. S. Perry and C. J. Van Oss (Eds.), Vol. 3 (New York: Wiley, 1970).
47. *Modern Plastics Encyclopedia.* (New York: McGraw Hill, 1970).
48. J. T. Baker Chemical Company. "Applications Bulletin on Baker-Flex Polyamide 6. Phillipsburg, New Jersey.
49. Scott, R. P. W. and P. Kucera. *Anal. Chem., 45,* 749 (1973).
50. Scott, R. P. W. and P. Kucera. *J. Chromatog. Sci., 11,* 83 (1973).
51. Data Sheet. *Incremental Gradient Elution Methodology.* (Clifton, N.J.: H. Reeve Angel and Co., Inc., 1973).

52. Zawta, B. and W. Holzel. *Pharmazie*, *23*, 174 (1968); and 236 (1968), 2 parts.
53. Soczewinski, E. and H. Szumilo. *J. Chromatog.*, *81*, 97 (1973).
54. Rabel, F. M. *Am. Lab.*, *6*, 33 (1974).

CHAPTER 8

BONDED PHASES ON SMALL POROUS PARTICLES

Ronald E. Majors

INTRODUCTION

Developments in Small Porous Packings

Porous adsorbents, such as silica gel and alumina, of large particle sizes (150 μm and greater) have been used for many years in liquid chromatography (LC). As the particle diameter (d_p) decreases, column efficiency should increase due to more rapid solute mass transfer between the shallower pores and smaller mobile phase channels between particles. Early investigations[1-3] found that H (or HETP, height equivalent to a theoretical plate) values of columns packed with irregularly shaped adsorbent particles decreased with a decrease in d_p until a d_p of about 40 μm was reached. Lack of suitable packing techniques as well as the general unavailability of smaller adsorbent particles in narrow size ranges for d_p of less than 40 μm hampered further development in the use of small porous particles.

To circumvent these problems, porous layer bead (PLB) adsorbents were developed and gained widespread acceptance. These particles have a thin porous adsorbent layer of 1 to 2 μm thickness on a 40-μm solid core glass bead. This thin layer permits solutes to diffuse rapidly into and out of the stationary phase. The dense, spherical particles could be easily dry-packed, gave columns with reasonably high efficiency, and could be used with the sensitive ultraviolet adsorption detector for analytical LC. However, due to the small amount of available surface area, PLB packings suffered from low sample capacity. They were often overloaded with resultant efficiency loss when attempts were made to inject the larger samples (0.05 to 0.5 mg) required for detectors of lower sensitivity, such as the refractive index detector. For preparative chromatography, large bore columns of PLB adsorbents were required for sufficient sample capacity, but at considerable expense.

Recent advances in the preparation and classification of porous particles in commercial quantities in the $5 \leq d_p \leq 20$-μm range and development of suitable techniques for packing such particles by high pressure slurry techniques[3-6] and special dry-tamping procedures[7] have renewed interest in porous particles for high performance LC. Systematic investigations of effects of d_p on column performance show[8] that 5- to 10-μm particles, properly packed, give efficiencies two orders of magnitude greater than 40-μm porous particles while retaining the major advantage of large sample capacity. In addition, porous packings with $d_p < 20$ μm offer greater efficiencies than PLB adsorbents.[9] However, to realize the full advantage of increased performance of small porous particles packed into long columns, high pressure pumps with output pressures in excess of 3000 psi are desirable.

Developments in Bonded Phases

Although, conceptually, any number of chemical species may be bonded to adsorbent particles, in practice thus far only three general types of chemical bonds have been used with LC packings. Since silica gel contains large numbers of silanol (≡ Si-OH) groups extending from the surface, the silanol has been esterified by reaction with alcohols to yield silicate esters[10] or silylated by reaction with organochlorosilanes[11-13] and organoethoxysilanes[14] to give siloxane bonds. Most commercial bonded phase packings are of these types. Although not commercialized, Si-C type bonded phases have been made by replacement of silica-OH radicals with chlorine followed by Grignard, Wurtz, or other reactions which convert the Si-Cl bonds to Si-C bonds.[15]

Silicate ester bonded phases have not proved entirely adequate, due to the lack of hydrolytic and thermal stability of the Si-O-C bond. Siloxane (Si-O-Si) bonded phases have proved more popular since a number of organosilane reactants are commercially available, the reactions are easier to carry out and, under controlled conditions, are reproducible

More importantly, the siloxane phase possesses sufficient stability under the conditions used in most chromatographic applications. These bonds will normally be attacked only in very acidic or basic environments. The stable Si-C bonded phases are less widely used since salt byproducts of the reactions may be occluded in the packing and are difficult to remove. This type of bonded phase may be the most chemically stable of the three discussed, provided preparation difficulties can be overcome. Conditions in which the silica base itself may be attacked (*e.g.*, in very basic aqueous solution) will still cause loss of this type bonded phase since silica is composed of Si-O-Si bonds. Since bonded phase characteristics may differ from conventional coated liquid-liquid chromatographic (LLC) phases, chromatography on chemically bonded phases is termed "Bonded Phase Chromatography" or BPC to distinguish it from LLC.

Most commercial BPC packings have utilized either large porous silica particles (d_p of 50 μm or greater) or PLB adsorbents as the base material. Due to their higher surface area and large numbers of silanols, the larger diameter porous particles normally have a large percentage of bonded phase (5 to 10 wt. %). They therefore have a high sample capacity, but due to the large particle size give poor column efficiencies. On the other hand, PLB bonded phases exhibit good efficiency, though the small surface area of the base material limits PLB to low surface coverages, typically 1 to 2% by weight. Sometimes, to bond a sufficient amount of stationary phase, polymerization of organosilanes must be carried out.[12-13] Highly cross-linked polymeric phases are undesirable since slow solute diffusion in the phase may result in lower chromatographic efficiency.

Combination of Small Porous Packings and Siloxane Bonded Phases

Advantages of (1) small porous particles with d_p < 10 μm in LC and (2) of bonded siloxane phases have been discussed. Combination of the two should give packings that exhibit both high efficiency and high capacity. A preliminary report of siloxane phases bonded to 10-μm silica gel confirmed this belief.[16] The present chapter discusses the properties and applications of several types of polar and nonpolar siloxane phases bonded to 10-μm silica gel.

EXPERIMENTAL

Apparatus

The apparatus used to pack columns by the high-pressure, balanced-density slurry procedure is reported elsewhere.[4] It consists of a high pressure pump, a reservoir to hold the slurry of bonded phase packing and valving to permit rapid transfer of the slurry into a stainless steel chromatographic column. A balanced density solvent is one that has a density equal to that of the BPC particles so there is a minimum of size segregation during the packing procedure.

Liquid chromatography was performed using a Varian Aerograph 4200 Research Liquid Chromatograph with a Multi-Linear Solvent Programmer (MLSP). This chromatograph employs two constant flow rate syringe pumps capable of output pressures up to 5000 psi. Mobile phases were custom-blended by automatic adjustment of the flow ratio of two solvents (A and B) each placed in a separate pump. The programmer permits the chromatograph to be used for either constant mobile phase composition for isocratic elution, solvent programming (gradient elution) or flow programming. Gradients were formed by continuous variation of the mobile phase composition at constant total pump displacement rates. Solvent programs were formed with the MLSP which permits formation of both simple linear and complex nonlinear positive or negative gradients. Nonlinear programs may be generated by a series of one to ten linear steps of selectable slopes and duration. Column regeneration, when required, was accomplished by either negative retracing of the positive gradient or quick reset to the initial mobile phase composition.

Injections were by the stop-flow technique made with a syringe onto a thin porous Teflon™ disc tightly pressed onto the top of the column. Injection directly into the packing eventually deteriorates column performance due to a disturbance of the packed bed by the fluid entering at high velocity from the syringe needle. A fixed wavelength (254 nm) ultraviolet adsorption detector, a refractive index monitor, or a variable wavelength Variscan™ spectrophotometric detector, all available from Varian, were used.

Reagents

Chromatographic solvents were of spectrophotometric grade available from Matheson, Coleman and Bell (Los Angeles, Calif.). A 50/50 V/V mixture of tetrabromoethane and tetrachloroethylene (both available from J. T. Baker, Phillipsburg, N.J.), their ratio being slightly adjusted for each type of bonded phase, was used as the balanced density solvent.

Solutes were the best grade available and were used as received. Antioxidants were provided courtesy of the following suppliers: Antioxidant 736, 754 (Ethyl Corp.), BHT butylated hydroxytoluene (Eastman Kodak), CAO-14 (Catalin Corp.), Irganox 1010, 1076 (Ciba-Geigy), and Goodrite (B. F. Goodrich).

Column Packings and Columns

LiChrosorb™ Si 60, a product of E. Merck (Darmstadt), was the base silica gel used for the bonded phases. Although available in a number of average particle sizes, only the ones designated by the manufacturer as 5, 10 and 30 µm were used. The phases were bonded by reaction of the LiChrosorb with organosilane reagents in dry toluene without intentional polymerization. After reaction, the packings were exhaustively extracted in a Soxhlet apparatus with benzene, acetone and methanol in that order. Most of the experiments described herein were carried out on MicroPak™ bonded phase columns. Inquiries regarding the availability of these prepacked columns containing 10-µm average diameter particles should be directed to Varian Instrument Division, 611 Hanson Way, Palo Alto, California 94303. MicroPak-CH refers to a nonpolar BPC column packed with 10-µm silica with octadecylsilane groups extending from its surface. The suffix -CH denotes the hydrocarbon functionality. MicroPak-CN and -NH_2 are polar bonded phase columns with cyanoalkyl and aminoalkyl functional groups, respectively.

In addition to the totally porous silica gel substrate, a PLB packing was used in preparing a bonded phase. Perisorb™ A, manufactured by E. Merck (Darmstadt), was used as the base silica and octadecylsilane groups were bonded to its surface. This BPC packing will be referred to as Perisorb-ODS. Permaphase™ -ODS (DuPont) is another PLB material which was studied. It is structurally similar to the Perisorb-ODS.

RESULTS AND DISCUSSION

Effect of Particle Diameter and Viscosity on Efficiency and Pressure Drop

For liquid-solid chromatography, earlier studies[9] had revealed a strong dependence of H on d_p. For silica and alumina and for solutes of k' value of about unity, H was proportional to $d_p{}^n$ where n was 1.8. For larger values of k', n changed only slightly although it was not found to be constant.[17,18] Similarly, the data in Figure 8.1 for a bonded phase packing of octadecylsilane groups on silica gel demonstrates the strong dependency of H on d_p. These results are in agreement with our earlier data[8,17] and appear to be a general trend of increased efficiency of small-diameter particles packed by the balanced density technique. For comparison to porous layer bead packings, other data points are shown for Perisorb-ODS and Permaphase-ODS using the same test solute and mobile phase. Although k' values were slightly different for each type of packing, from an efficiency standpoint, bonded phase porous packings of d_p less than about 20 μm were equivalent to or superior to bonded phase PLB material of d_p approximately 40 μm. These observations are in agreement with earlier data for unbonded small particle silica and alumina.[9]

Smaller particles require higher column head pressure than larger particles when packed into columns of equal dimensions and operated under equivalent conditions. Typically, a 15-cm x 2.2-mm column packed with nominal 10-μm porous particles gives a backpressure of about 350 psi at 20°C when operated at 2 ml/min (linear velocity = 1 cm/sec) with a mobile phase of hexane (η = 0.3 centipoise). For higher viscosity solvents such as those used in reverse-phase chromatography (*e.g.*, water and isopropanol) pressure drops are greater, being proportional to mobile phase viscosity. An earlier study[8] revealed that, for slurry-packed porous particle columns, each time the particle diameter was halved column backpressure increased 3.7 ± 0.1 times.

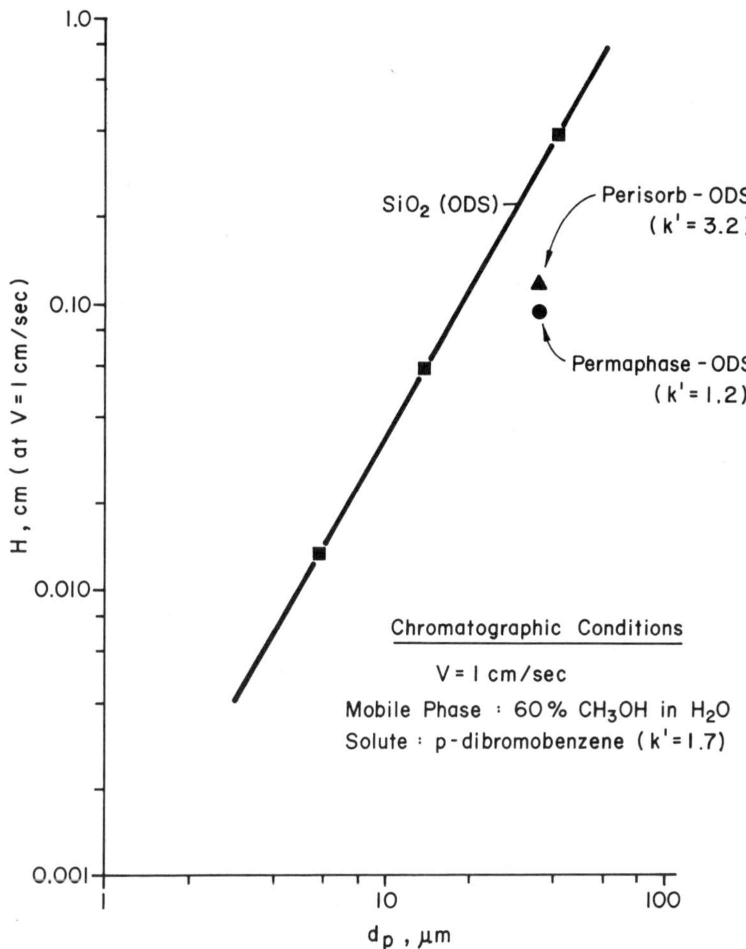

Figure 8.1. Dependence of HETP on particle diameter for hydrophobic octadecylsilane chemically bonded phases.

Chromatographically Significant Properties of Bonded Phase Columns

Siloxane phases are resistant to stationary phase "bleed," as normally experienced in liquid-liquid partition chromatography (LLC). Thus, unlike LLC, no presaturator is required to ensure mobile phase-stationary phase equilibrium. The solute may be collected in the column effluent uncontaminated with liquid phase from the column, mobile phases of high solvent strength may be used, and most importantly solvent programming (gradient elution) may be used for samples with widely varying k' values. A recent review of bonded phases by Locke[19] discusses these points in detail.

Effect of Temperature on Retention, Efficiency and Pressure Drop

The effect of temperature on the retention characteristics of MicroPak-CH for three test solutes using a 60% methanol in water mobile phase is depicted in Figure 8.2. For each solute, a straight line relationship exists between log k' and $1/T$.

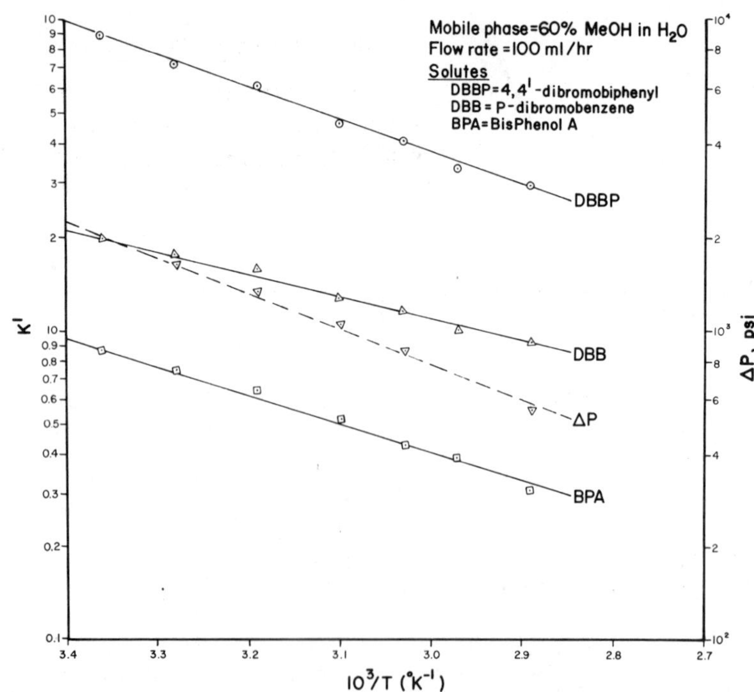

Figure 8.2. Effect of temperature on capacity factor and column pressure for MicroPak-CH.

Increased temperature generally increases solute solubility in the mobile phase while decreasing solute stationary phase interactions. On the average, a 35°C increase in temperature resulted in a decrease in retention by a factor of 2. This behavior is in agreement with data from other reverse phase packings.[20,21] The slope of each plot is slightly different and indicates that, in addition to retention, selectivity (relative retention) may also be altered by a change in temperature.

Furthermore, it can be seen in Figure 8.3 that a higher column temperature improved chromatographic efficiency probably due to a reduction of mobile phase viscosity and resultant increase in the solute diffusion rate. The *H* values were decreased by about

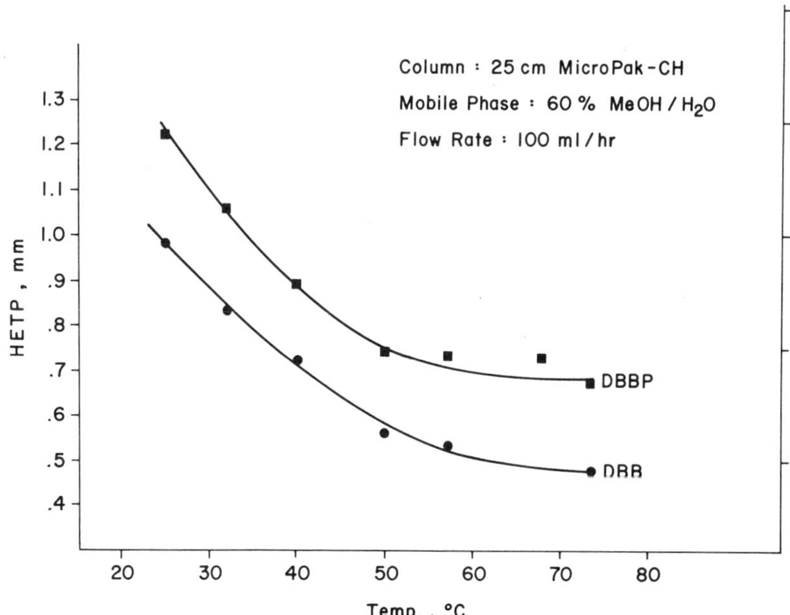

Figure 8.3. Relationship between efficiency and temperature for MicroPak-CH; same solutes as in Figure 9.2.

40% for an increase in column temperature from 25°C to 50°C. The latter temperature appears to be optimal, for beyond 50°C the incremental gain in efficiency may be offset by problems of solvent outgassing. Finally, the decrease in mobile phase viscosity with temperature also resulted in a decrease in column head pressure as Figure 8.2

indicates. Thus, for optimum performance with the viscous mobile phases frequently encountered in reverse-phase operation, elevated temperatures are recommended.

Effect of Mobile Phase Concentration

Although temperature may be used as a variable to change k' values, the results in Figures 8.4 and 8.5 suggest that a variation in the mobile phase concentration has a more powerful influence on

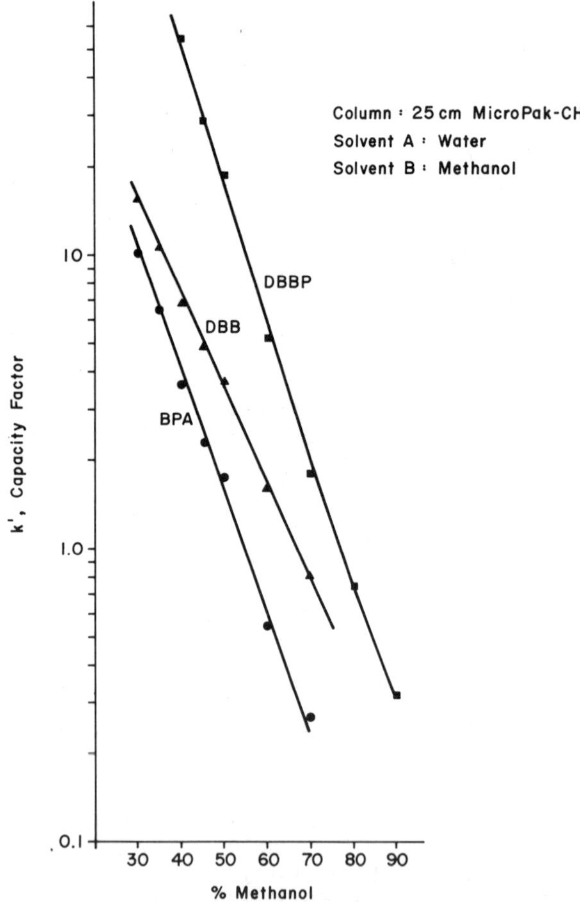

Figure 8.4. Capacity factor as a function of methanol concentration for MicroPak-CH; temperature: 50°C.

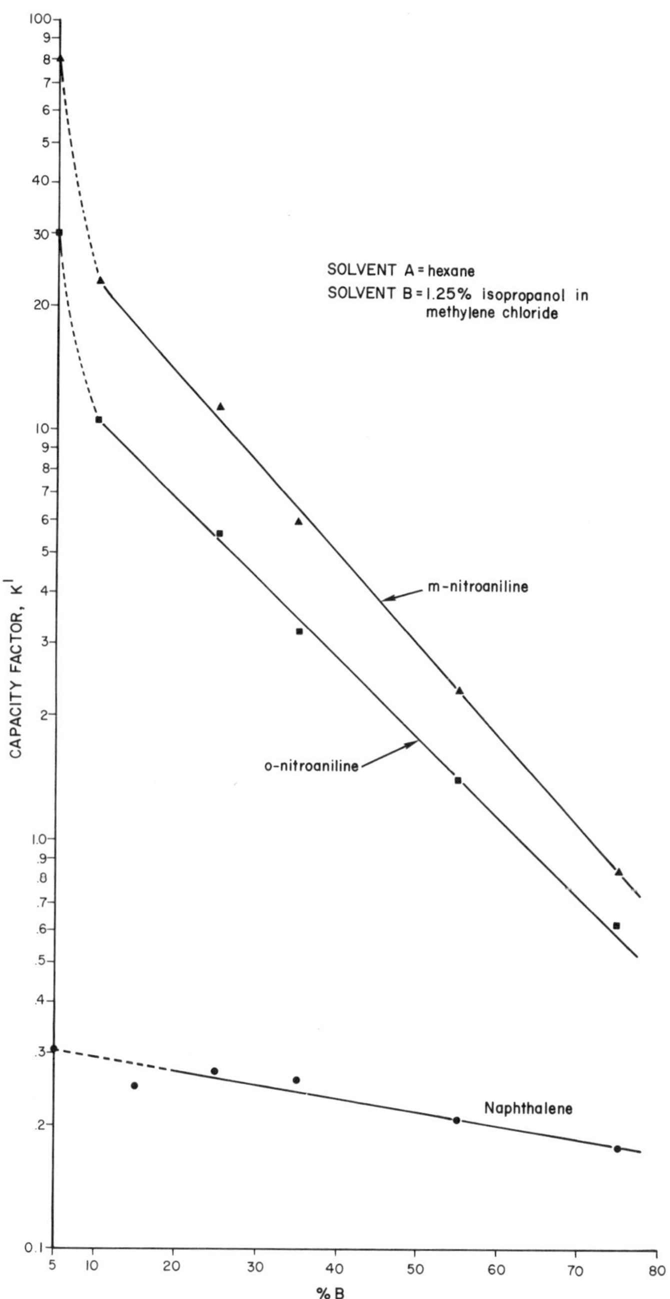

Figure 8.5. Capacity factor as a function of mobile phase composition for MicroPak-NH_2; temperature: ambient.

retention. For reverse-phase chromatography at 50°C using MicroPak-CH, Figure 8.4 shows that for the three test solutes a 10% increase in the concentration of methanol in water gave about a 2½ to 3-fold decrease in k' values. The k' values of the components showed different concentration dependencies maybe due to varying degrees of solvation among the polar and nonpolar species. Other water-soluble organic modifiers with ultraviolet transparency such as dioxane, acetonitrile, or isopropanol may also be used to affect k' values. Increasing the organic modifier concentration with time is a convenient way of performing gradient elution for the separation of samples containing weakly and strongly retained components. Examples of such separations will be shown later.

A similar plot of log k' versus per cent modifier for the polar bonded phase column MicroPak-NH_2 is depicted in Figure 8.5. At ambient temperature, mobile phases consisting of hexane (nonpolar solvent A) with increasing amounts of 1.25% isopropanol in methylene chloride (more polar solvent B) were custom-blended using the solvent programmer in the isocratic mode. For the weakly retained component naphthalene, there was little change in k' with the concentration of solvent B. However, above 10% B a straight line relationship existed between log k' and per cent B for the polar test solutes, *o*- and *m*-nitroanilines. For those compounds, a 10% increase in the concentration of solvent B resulted in a 30-40% decrease in k' values, somewhat less change than observed for the MicroPak-CH example. This difference may possibly be explained by the great difference in polarity between the hydrophobic octadecyl phase of the MicroPak-CH and the aqueous methanol mobile phase compared to the polarity between alkyl amine phase of MicroPak-NH_2 and the hexane-methylene chloride-isopropanol mobile phase. Interestingly, for the polar solutes, the data points at 5% B, consisting of 95% hexane — 4.9% methylene chloride — 0.068% isopropanol, produced a distinct positive deviation from the straight line plot. This observation is consistent with our earlier work[16] and the work of others[12,13,22,23] where low polarity mobile phases appeared to give poor "wetting" (limited solvation) of polar bonded phases. When using polar bonded phases and nonpolar mobile phases, a polar modifier such as an alcohol 0.1% or greater by volume is recommended.

Dependence of Solute Retention on Time with Change in Mobile Phase Composition

Except in those cases as mentioned above where poor solvation of the bonded phase may occur, in contrast to liquid-solid chromatography, the stationary phase activity in BPC is not so critical and careful control of the modifier content of the mobile phase is not generally required. At the conclusion of a solvent program, BPC column may be returned quickly to the initial conditions merely by flushing out the higher strength mobile phase from the column. To demonstrate the rate at which bonded phases adjust to mobile phase composition changes, columns of MicroPak-CH and $-NH_2$ were subjected to rapid changes in mobile phase concentration and the k' values of a solute monitored by injections onto the column after each elution of the peak. After a step change in the concentration, a minimum of four to five minutes was allowed to flush the previous solvent from the column. The linear velocity was 1 cm/sec in all cases.

For MicroPak-CH, three step changes from higher to lower methanol concentrations were made and the k' value for 4,4'-dibromobiphenyl determined as a function of time as depicted in Table 8.1. In all three cases, k' values were constant within fifteen minutes and for the higher methanol concentrations within six minutes.

For MicroPak-NH_2 it can be seen in Table 8.1 that a step change in the concentration of more polar solvent gave a constant k' value within ten minutes for a solute of k' 10.2 ± 0.1. During a solvent programmed (gradient elution) run, the steroid progesterone showed k' value of 4.2 ± 0.1 for seven successive runs if, at the conclusion of the previous gradient run, the mobile phase was immediately returned to the initial concentration and that concentration passed through the column for five minutes prior to injection and start of the next solvent program run. This experiment indicated that regeneration versus time is repeatable but did not necessarily prove that it was complete. Further work is in progress on regeneration rates of BPC columns.

Table 8.1

Dependence of Solute Retention on Time with Change in Mobile Phase Composition

Column	*Solvent A*	*Solvent B*	*Concentration Change*	*Type of Change*	*Time, min.*[a]	*k'*	*Solute*
MicroPak-CH 25 cm x 2.2 mm	Water	Methanol	80% to 70%	Step	6	2.2	4,4' dibromobiphenyl
					15	2.1	
					19	2.2	
					22	2.1	
					335	2.1	
			70% to 60%	Step	4.5	5.9	
					9	6.0	
					12	5.9	
			60% to 50%	Step	4	21.4	
					15	20.4	
					25	20.1	
MicroPak-NH_2 15 cm x 2.2 mm	Hexane	1.25% IPA[b] in CH_2Cl_2	20% to 10%	Step	10	10.3	*o*-nitroaniline
					35	10.1	
	0.13% IPA in hexane	50% IPA in CH_2Cl_2	2% to 30% to 2%	Gradient, then step back to initial	5	4.2 ± 0.1[c]	progesterone

[a]Linear velocity, 1 cm/sec

[b]IPA = isopropanol

[c]Reproducibility of *k'* values during gradient after five minute flush with 2% B. Average deviation of 7 runs.

Sample Capacity Effect

Compared to porous layer BPC packings, the higher coverage of the bonded stationary phases on porous particles allows larger amounts of sample to be injected onto the column. Studies of linear capacity of a packing give an idea of the quantity of sample which may be injected before overload occurs with subsequent loss of resolution. Linear capacity is determined by successively larger injections of a solute until there is a significant increase in H value.[24] It is typically expressed in milligrams of solute per gram of packing. For a 25-cm by 2.2-mm i.d. MicroPak-CN column, such a study was performed at two temperatures using two nitroaniline isomers as solutes and 40% methylene chloride in hexane containing 0.5% isopropyl alcohol as the mobile phase. The results in Figure 8.6 show that at 25°C

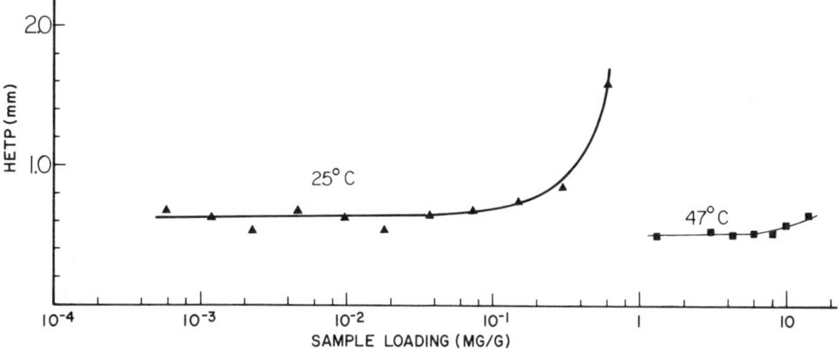

Figure 8.6. Effect of sample loading on column efficiency for MicroPak-CN at two temperatures. Solute: Solute: o-nitroaniline.

for *o*-nitroaniline (k' = 1.2) column overloading begins to occur at 0.3 mg/g and becomes significant at 0.5 mg/g. Although not shown, for *m*-nitroaniline (k' = 2.5) overloading occurred at the slightly higher value of 0.7 mg/g when the column was operated at 25°C.

Increase in column temperature to 47°C had a dramatic effect on sample capacity for *o*-nitroaniline as depicted in Figure 8.6. In addition to a lower value of H, raising the temperature to 47°C permitted 15 mg/g to be injected before a noticeable loss in efficiency occurred. That limit was governed by the solubility of the solid *o*-nitroaniline in the injection solvent, isopropanol. For the 25-cm by 2.2-mm column, that represented an injection of about 7.5 mg. For *m*-nitroaniline, a sample size of 2 mg was injected at 47°C, again the limit being caused by injection solvent solubility rather than by column capacity.

Quantitative Ability of Bonded Phase Columns

For quantitative analysis, both column and detector linearity are important. To illustrate the quantitative ability of a bonded phase column-detector system, a calibration curve was measured for *m*-nitroaniline on a 25-cm MicroPak-CN column. A Variscan detector was set at 254 nm for these measurements. The detector response as a function of the number of micrograms injected as indicated in Figure 8.7. Both peak height and peak area

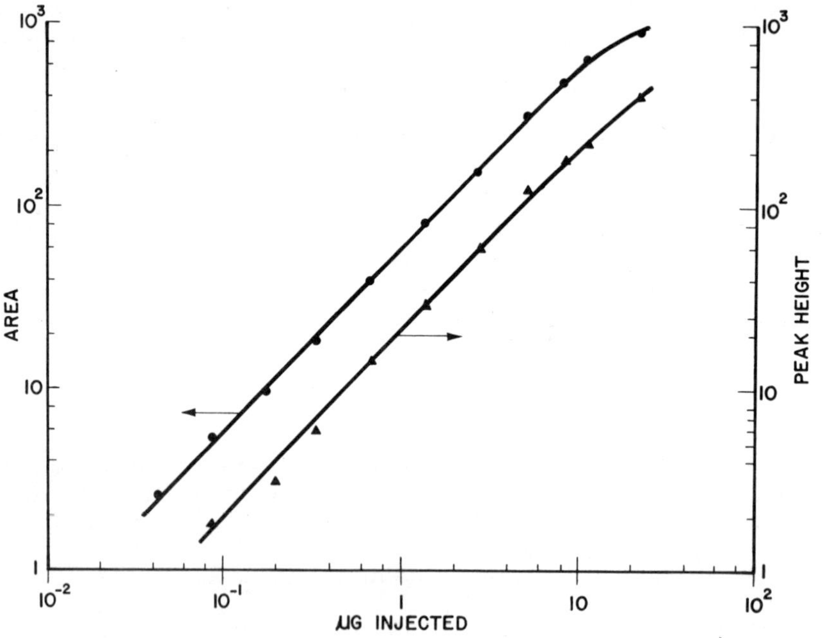

Figure 8.7. Calibration curve for m-nitroaniline on MicroPak-CN; detector wavelength: 254-nm; mobile phase: 65% hexane - 34.5% CH_2Cl_2-0.45% isopropanol.

(measured by triangulation) expressed in arbitrary units gave a straight line plot with the expected slope of one. There was less scatter observed for peak area measurements. The nonlinearity at the highest sample concentration was due to expected deviations from Beer's law. Peaks developed a square top for peak heights measured about 2.7 absorbance units.

APPLICATIONS OF BONDED PHASE COLUMNS

MicroPak-CH

Compounds only slightly retained or unretained by normal LLC, LSC, or BPC (*i.e.*, polar stationary phase, nonpolar mobile phase) are best separated by reverse-phase chromatography. MicroPak-CH possesses a hydrophobic surface and is most useful for reverse-phase BPC. As a rule of thumb, in reverse-phase BPC polar solutes elute first and retention times increase with decreasing solute polarity. Most frequently used mobile phases with MicroPak-CH are distilled water in combination with water-miscible organic solvents such as alcohols (methanol, ethanol, isopropanol), acetonitrile, dioxane and tetrahydrofuran. As was noted in Figure 8.4, increase in the amount of organic modifier has the effect of decreasing solute retention.

A typical reverse-phase isocratic separation of alkyl bromides at 50°C is presented in Figure 8.8.

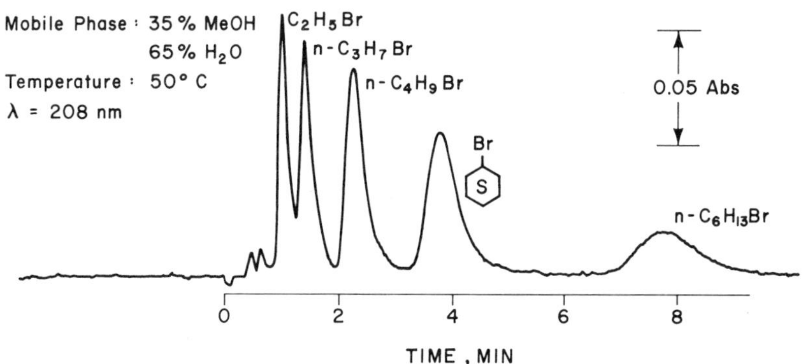

Figure 8.8. Separation of alkyl bromides on MicroPak-CH using Variscan at 208 nm.

At the 254-nm wavelength of most single-wavelength UV detectors, such compounds gave no absorbance. However, the bromide group gives rise to a moderately strong electronic transition at 208 nm and the Variscan was set at this wavelength. For cyclohexyl bromide, the detector response represented about 40 µg. Note that the most polar C_2H_5Br eluted first, and as solute polarity decreased retention time increased. In order to elute the less polar decyl and dodecyl bromides in a reasonable amount of time, a water-methanol solvent program was required. Such a separation can be seen in Figure 8.9a. Unfortunately due to the UV cutoff for methanol at 205 nm,[25] very close to the wavelength setting of 208 nm, an appreciable baseline shift occurred. A baseline obtained in the absence of sample is shown in Figure 8.9b.

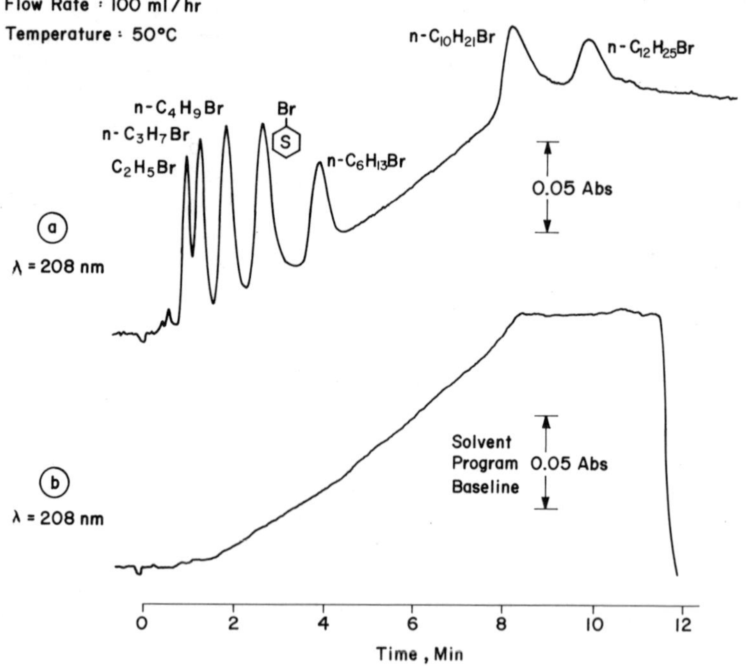

Figure 8.9. (a) Solvent programmed separation of alkyl bromides using Variscan.
(b) Solvent program baseline; run in the absence of sample; baseline change due to absorbance of methanol at 208 nm.

Sterically-hindered phenols are used widely as effective nondiscoloring, nonstaining antioxidants for polymer systems, particularly rubbers and plastics. They are added in 0.05-0.2% by weight and often, for best results, several are used in combination. Since the phenolic group is hindered by bulky substituents, the compounds exhibit low polarity and are separated by virtue of their hydrocarbon character. A linear solvent-programmed separation of several antioxidants used in polyolefins is depicted in Figure 8.10. The elution order

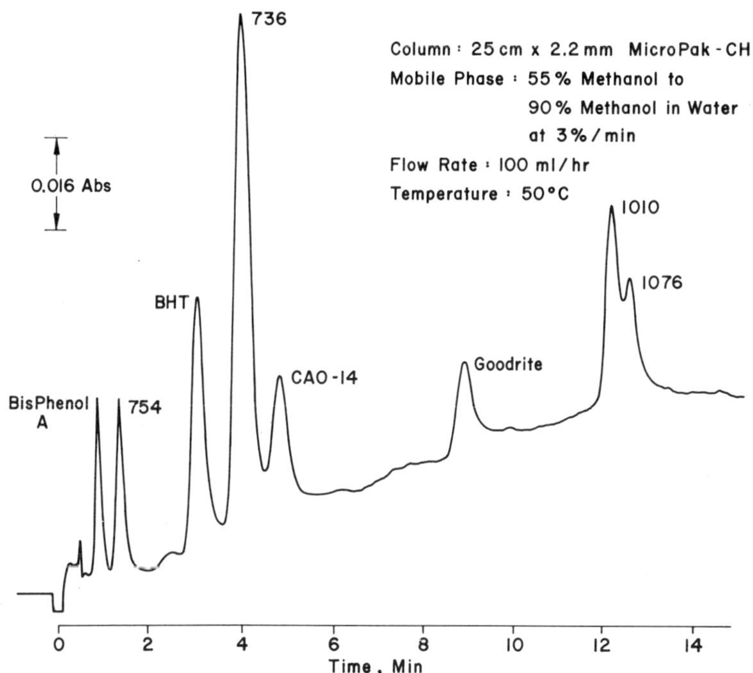

Figure 8.10. Solvent programmed separation of hindered phenolic antioxidants on MicroPak-CH. Solutes identified under Experimental.

is reversed for some compounds when separated by LSC on alumina.[16] Attempts to separate Irganox compounds 1010 and 1076 by LSC or normal BPC met with failure probably because of their predominantly hydrocarbon character with lack of differing functionality. With a longer reverse-phase column and less rapid gradient increase, the two could have been separated to baseline but at the expense of separation time.

Plasticizers are added to polymers to impart a desirable degree of flexibility over a broad range of temperature and to lower the brittle point. Plasticizers must have good polymer solvent power, compatibility and low volatility. Their high boiling points make liquid chromatography a favorable technique for plasticizer analysis. Long-chain phthalates are frequently used as plasticizers for several polymeric systems such as polyvinylchlorides, polyacrylates and some cellulose materials. The reverse-phase separation of several 1,2 substituted phthalates of increasing alkyl chain length is demonstrated in Figure 8.11. The most polar, DMP, eluted first followed by those of increasing hydrophobic character with the C_{10}-substituted phthalate eluting last. Again the water-methanol gradient proved to be a versatile mobile phase system to effect adequate sample resolution in a short time with the MicroPak-CH column.

MicroPak-CN

The popularity of β, β'-oxydipropionitrile as a coated liquid phase in LLC suggested that a cyano alkylsilane bonded phase might offer some of the properties of the former. MicroPak-CN columns are useful for normal BPC (*i.e.*, polar stationary phase, nonpolar mobile phase). The columns are slightly less retentive than columns of silica gel and, in comparison, offer modified selectivity. A simple separation of three of the phthalate plasticizers on MicroPak-CN is presented in Figure 8.12. The elution order is reversed for the same compounds shown in Figure 8.11. Elution order reversal is sometimes useful when a minor component of interest occurs on the "tail" of a major component. By going to the opposite mode of separation the minor component can be made to elute first, hopefully with less interference from the major component which should elute later. In addition, the MicroPak-CN exhibited similar selectivity to Durapak-OPN,[26] a silicate ester nitrile phase.

The polar bonded phase should prove useful for solvent-programmed work. To illustrate the influence of gradient shape on the separation characteristics of MicroPak-CN, a number of closely related azo compounds were separated using solvent programs generated by the MLSP. Using a linear gradient the

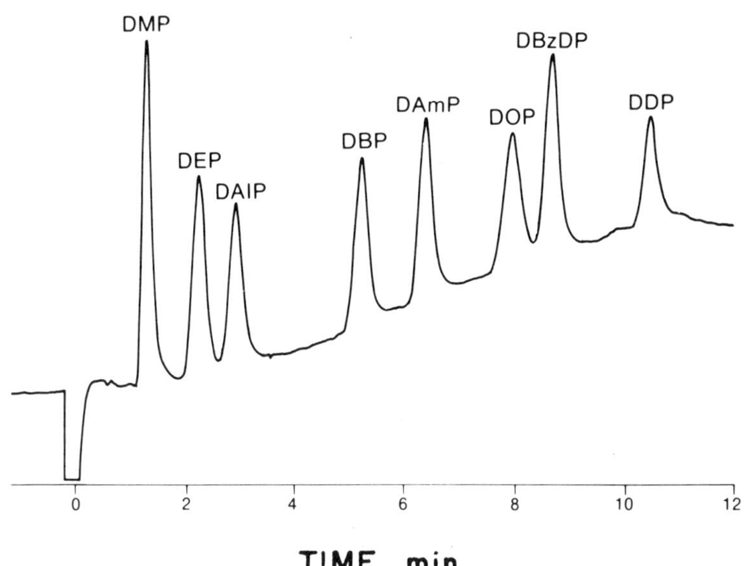

Figure 8.11. *Solvent programmed separation of phthalate plasticizers on MicroPak-CH.*
Column: 25 cm x 2.2 mm
Mobile phase: Solvent A, water. Solvent B, methanol, 40% methanol to 90% methanol at 5%/minute
Column temperature: 50°C
Detector: UV, 254-nm.

Sample: DMP – dimethylphthalate
DEP = diethylphthalate
DAIP = diallylphthalate
DBP = dibutylphthalate
DAmP = diamylphthalate
DOP = dioctylphthalate
DBzBP = benzylisodecylphthalate
DDP = didecylphthalate

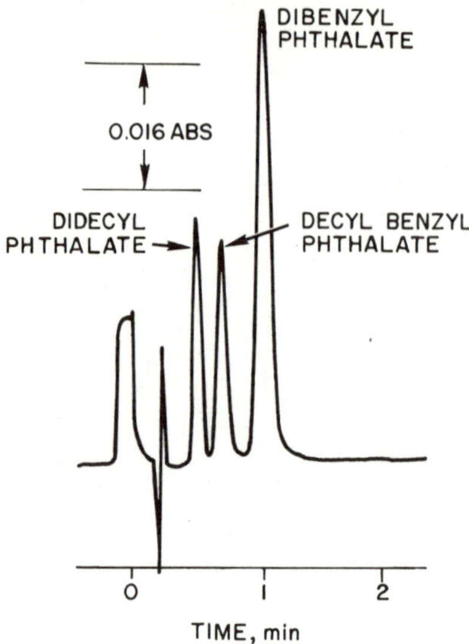

Figure 8.12. Separation of phthalate plasticizers on MicroPak-CN.[16]

partial separation of substituted *N,N*-dimethylazoanilines (structures included) is depicted in Figure 8.13. Note the partial resolution of peaks 2-3, 5-6 and 7-8. Application of a gradient delay followed by a nonlinear steeply rising gradient as shown in Figure 8.14 gave almost complete resolution of peaks 1 through 6, but 7 and 8 are still incompletely resolved. Components 7 and 8 might be expected to be difficult to separate due to the similarity of the substituent groups on their β-rings. A slower gradient rise or more preferably an isocratic "hold" prior to their elution could have resolved them better but at the expense of time.

MicroPak-NH_2

The adsorption behavior of silica and alumina is governed by the characteristics of the surface hydroxyl group. This group is generally regarded as slightly acidic due to its partial ionization.[24] Incorporation of the basic amino group into a chemically bonded phase should yield a packing with unique chromatographic selectivity. The amine functionality can interact with solutes by hydrogen bonding. It may, in fact, function as either a

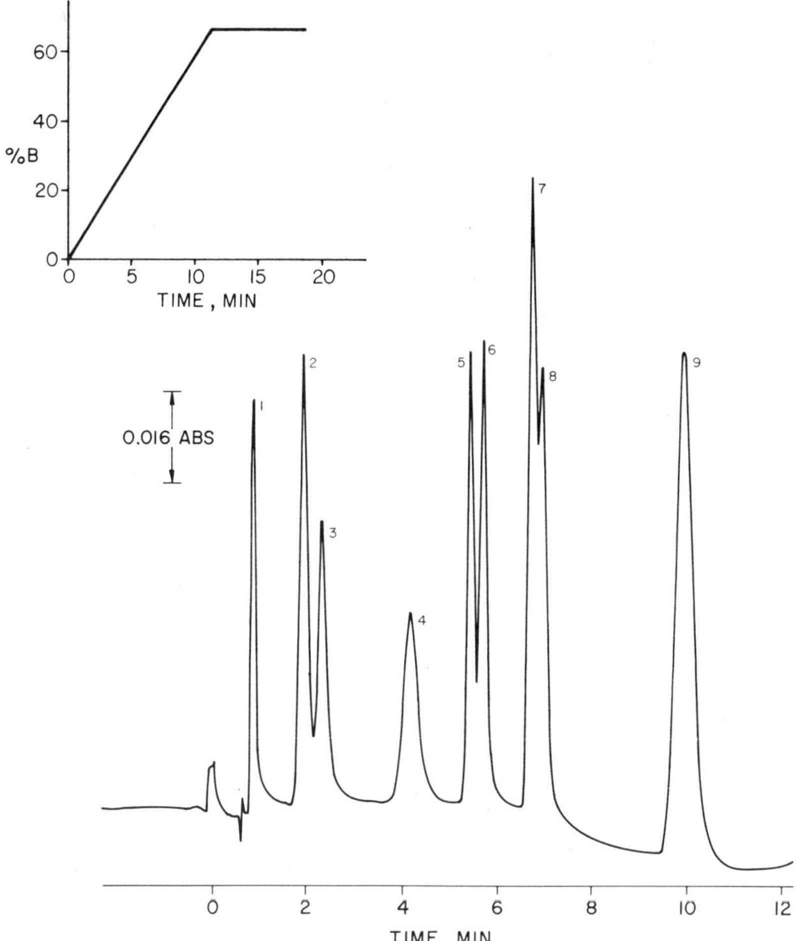

Figure 8.13. (a) *Solvent programmed separation of azo compounds on MicroPak-CN.*
Column: MicroPak-CN, 15 cm x 2.1 mm
Mobile phase: Solvent A, hexanes containing 0.2% isopropanol; Solvent B, methylene chloride containing 0.2% isopropanol;
Solvent program: shown on chromatogram
Flow rate: 60 ml/hr
Sample concentration: 0.3 mg/ml each except 0.09 mg/ml of 9
Sample size: 4 µl.

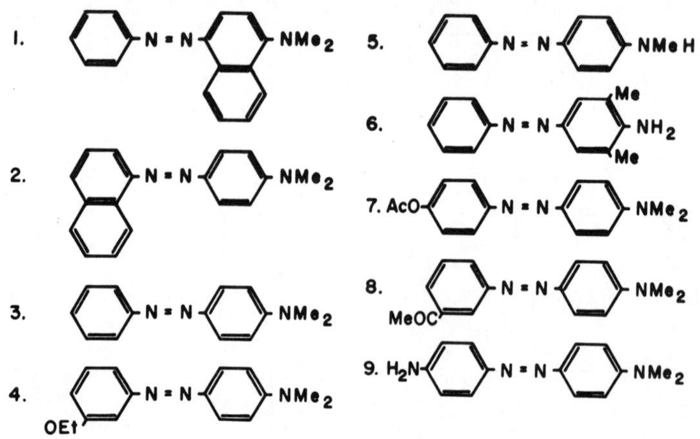

Figure 8.13. *(b) Structures of azo compounds used in Figures 8.13(a) and 8.14.*[16]

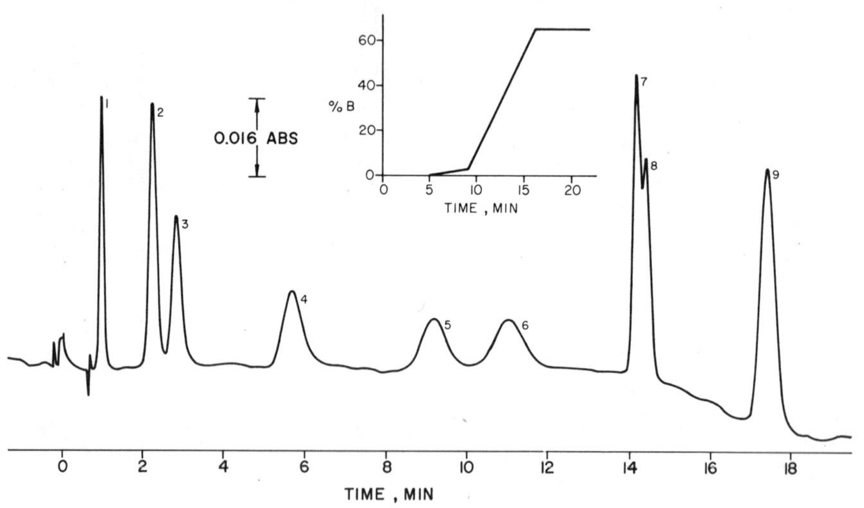

Figure 8.14. *Nonlinear solvent programmed separation of azo compounds on MicroPak-CN.*[16]
Chromatographic conditions: same as in Figure 8.13(a) except gradient profile changed as shown on figure.

Bronsted acid or base depending on the solute. In aqueous solution the amine may be protonated and serve as a potential anion exchanger.

The rapid separation at ambient temperature of adenosine mono-, di-, and tri-phosphate nucleotides (AMP, ADP, ATP) using 0.4 *M* KH_2PO_4 containing 30% methanol is illustrated in Figure 8.15. The methanol was included to sharpen the peaks. A MicroPak Si-10 silica gel column operated under the same conditions failed to resolve the three components. Despite the relatively viscous mobile phase ($\eta = 1.4$ cp determined by viscometry), the 15-cm column exhibited good efficiency. At a higher temperature, mobile phase viscosity would have been reduced and efficiency increased even further. The same separation has been carried out on a pellicular type of strong anion exchange resin[26] but required a much longer column.

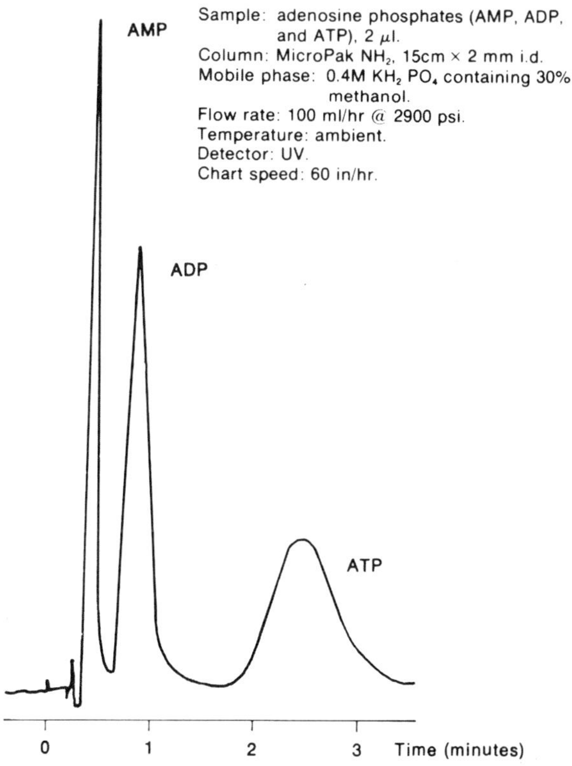

Figure 8.15. Separation of adenosine nucleotides on MicroPak-NH_2.

To further illustrate the advantage of variable gradient shape to effect optimum resolution within a reasonable time, eight closely related steroids of varying polarity were separated by solvent programming. Since some of the steroids were quite polar, a strong mobile phase consisting of equal volumes of methylene chloride and isopropanol was selected as the "B" solvent. First, a convex-shaped gradient whose profile is shown in Figure 8.16 was used. The chromatogram also illustrated in Figure 8.16 showed only partial resolution of the early eluting steroids. Obviously, for better resolution, a lower slope during the initial part of the gradient would be required. Sufficient resolution was obtained for the last three steroids.

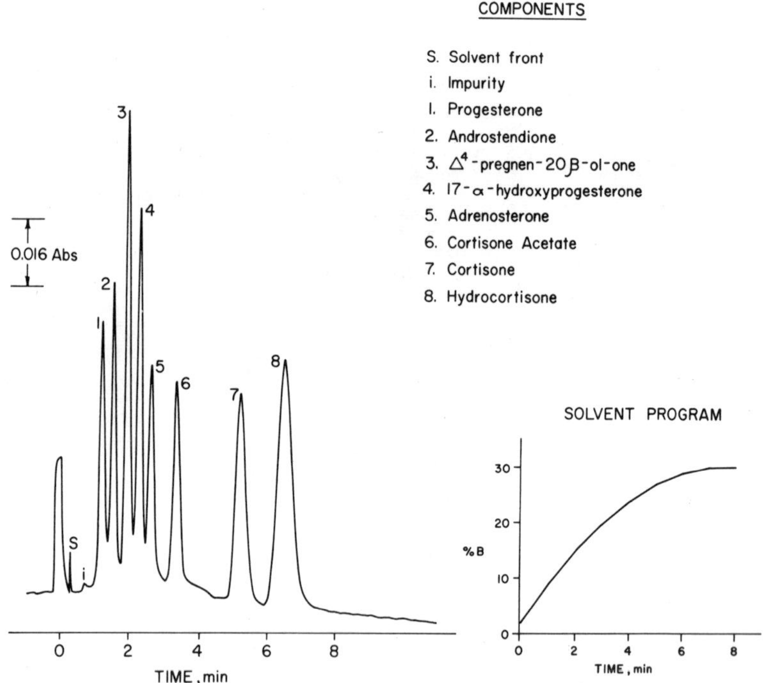

Figure 8.16. *Concave-shaped solvent programmed steroid separation on MicroPak-NH_2.*
Column: 15 cm x 2.2 mm
Mobile phase: Solvent A, hexane containing 0.2% isopropanol; Solvent B, 50/50 methylene chloride-isopropanol
Solvent program: shown on chromatogram
Sample size: 1.5 μg of each.

A concave-shaped solvent program composed of three linear steps gave good separation on the front end of the chromatogram while maintaining adequate resolution for the later eluting peaks as indicated in Figure 8.17. As a final modification, resolution

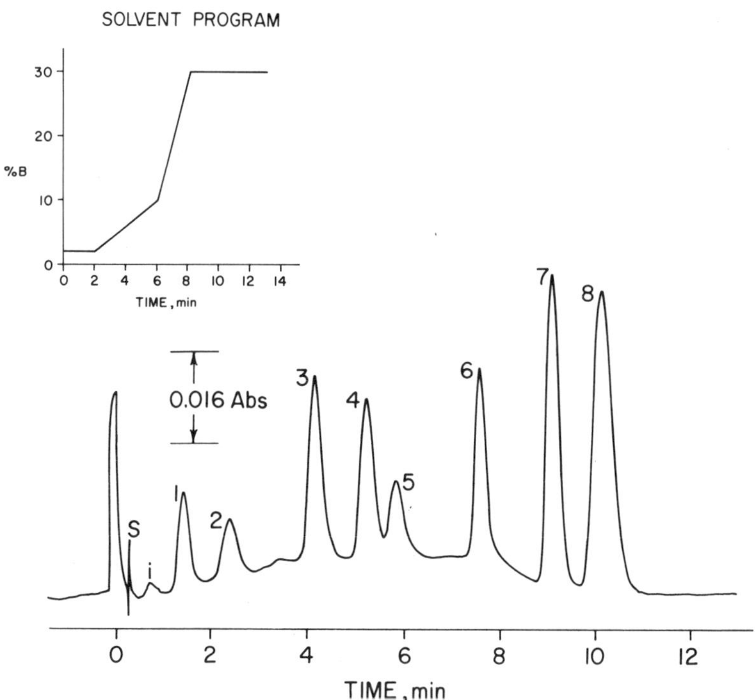

Figure 8.17. Convex-shaped solvent programmed steroid separation on MicroPak-NH_2.
Conditions: same as in Figure 8.16 except for gradient profile.

between peaks 4 and 5 was improved by providing an isocratic "gradient hold" for the two minutes prior to their elution. The chromatogram presented in Figure 8.18 shows all peaks completely resolved in less than 12 minutes. These chromatograms illustrate how a separation can be improved without a sacrifice in analysis time by systematically changing the solvent program.

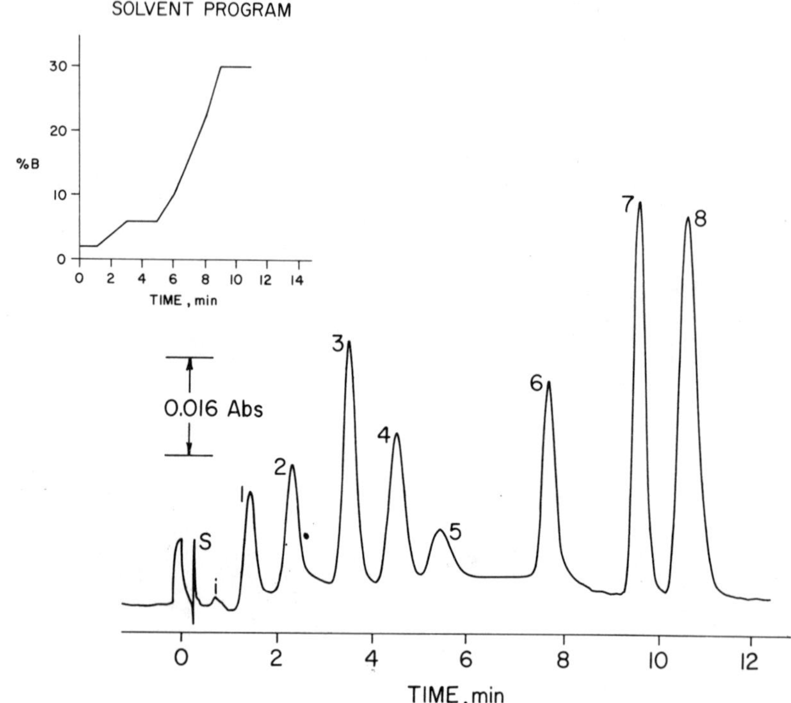

Figure 8.18. *Solvent programmed steroid separation on MicroPak-NH_2 using gradient hold.*
Conditions: same as in Figure 8.16 except for gradient profile.

Comparison of the Steroid Separation on Four Polar Columns

In LC, the selectivity is governed mainly by the stationary phase but also somewhat by the mobile phase. Relative elution order is determined by the affinity of the solutes for the functionality provided by the bonded phases. To compare the selectivity of the polar MicroPak columns, the eight steroids used in the previous analyses were separated on different columns using the same linear solvent program. Indeed both selectivity and retention may be appreciably altered by selection of a different bonded phase column as demonstrated in Figures 8.19

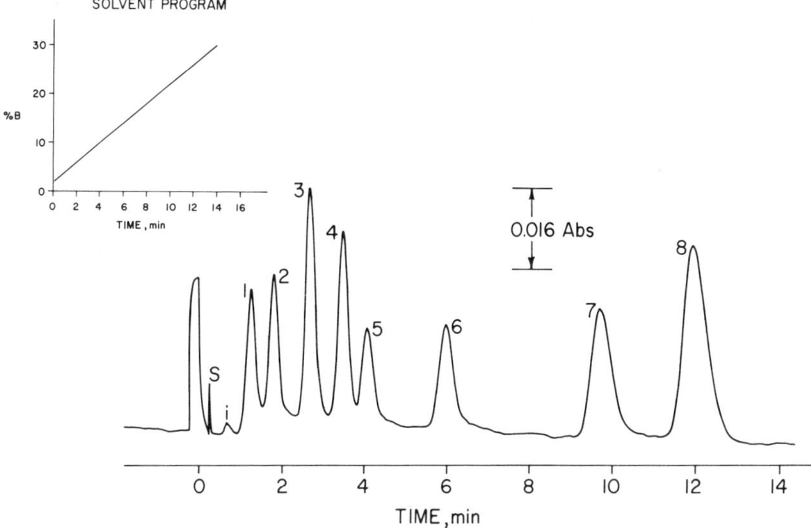

Figure 8.19. Linear solvent programmed steroid separation on MicroPak-NH_2. Conditions: same as in Figure 8.16 except for gradient profile.

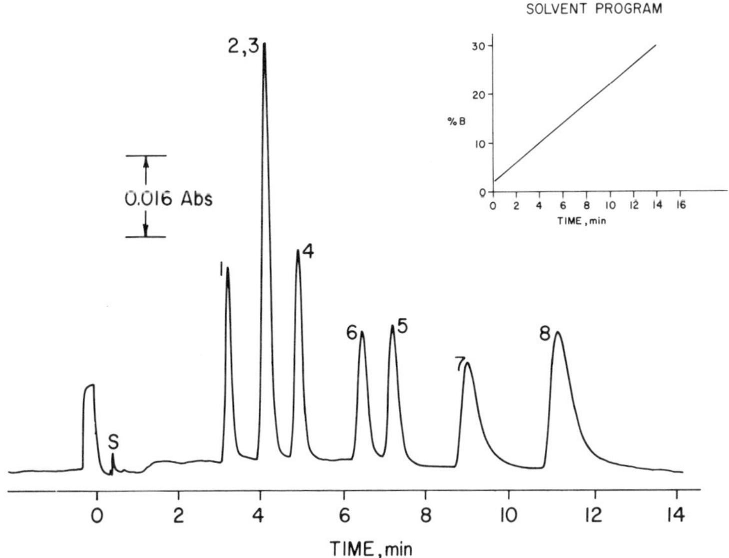

Figure 8.20. Linear solvent programmed steroid separation on MicroPak Si-10. Conditions: same as in Figure 8.19 except column is 15 cm x 2.2 mm. MicroPak Si-10 containing 10 µm silica gel.

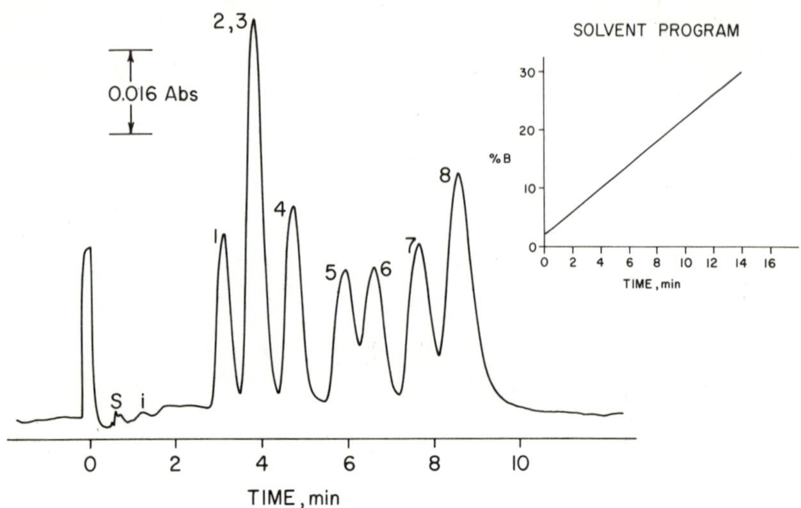

Figure 8.21. Linear solvent programmed steroid separation on MicroPak-CN. Conditions: same as in Figure 8.19 except column is 15 cm x 2.2 mm MicroPak-CN.

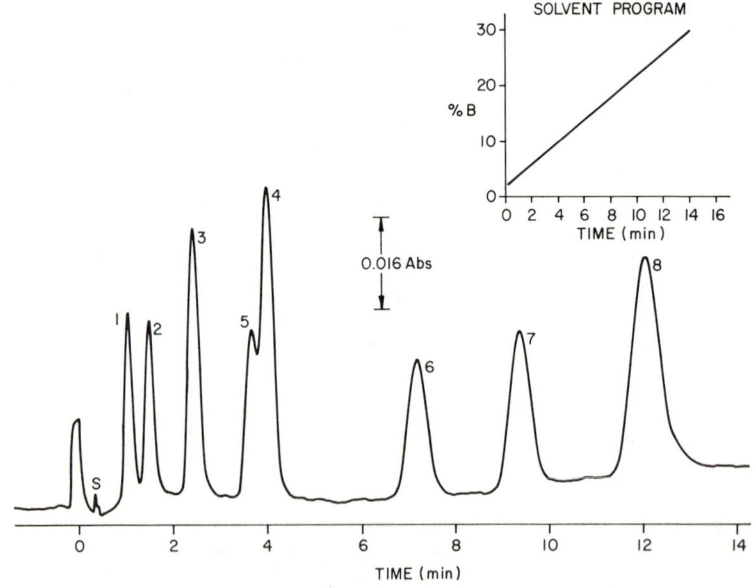

Figure 8.22. Linear solvent programmed steroid on pyridyl bonded phase column. Conditions: same as in Figure 8.19 except column is 15 cm x 2.2 mm containing a 2-pyridyl ethyl silane bonded to 10 µm silica.

to 8.22. The almost complete separation of the eight compounds on MicroPak-NH_2 is shown in Figure 8.19. Substitution of a column filled with 10-μm silica gel while keeping all other experimental conditions the same indicates that, although good efficiency is provided as evidenced by the narrow peaks, no separation is effected on steroids 2 and 3 (Figure 8.20). Furthermore, steroids 5 and 6 have reversed their elution order. By replacement with a MicroPak-CN column, the chromatogram in Figure 8.21 shows still lower resolution but the same elution order as on the MicroPak-NH_2 column. Finally, Figure 8.22 demonstrates further selectivity alteration by use of an experimental 2-pyridyl ethyl siloxane phase bonded to 10-μm silica gel. In this case, when compared to Figure 8.19, steroids 4 and 5 are reversed and only partially resolved. Application of a gradient hold prior to the unresolved doublet would have further separated those compounds. The last four examples illustrate the potential advantages of having several columns with varying degrees of selectivity available and how choice of the right one may more easily solve a separations problem. Remember that in addition to modifying retention, manipulation of the mobile phase may affect separation order as well.

CONCLUSIONS

Siloxane phases chemically bonded to small porous particles have a definite place on the liquid chromatographer's shelf. Both polar and nonpolar columns should be available. Each offers unique selectivity, important for different separations. The bonded phase packings have high linear sample capacity in the range of several mg/g, especially at elevated temperatures. They respond quickly to step or continuous changes in mobile phase composition, often requiring only several minutes to stabilize and give reproducible retention times. While capacity factors may be altered by changes in column temperature, they are more strongly influenced by variation of mobile phase composition. For best efficiency, elevated temperatures and use of particles of less than 10 μm in diameter are preferred. Elevated temperatures also reduce pressure requirement by decreasing mobile phase viscosity. Bonded phases may be used in wide areas of applications. Due to the stability of the siloxane bond, they are quite useful in solvent programming.

In the future, due to their obvious advantages, chromatography on the bonded phases may likely obsolete the technique of conventional liquid-liquid chromatography. Hopefully, unlike gas chromatography with its hundreds of stationary phases, due to the ability to manipulate the character of the mobile phase, only a few (less than 15 to 20) bonded stationary phase packings will be required to cover most of the separations.

ACKNOWLEDGMENTS

The experimental assistance of Margaret Hawken and Manual Madruga is gratefully acknowledged. The author also thanks several of his colleagues at Varian for their helpful comments on the manuscript.

REFERENCES

1. Snyder, L. R. "An Experimental Study of Column Efficiency in Liquid-Solid Adsorption Chromatography," *Anal. Chem.*, *39*, 698 (1967).
2. Stewart, H. N. M., R. Amos, and S. G. Perry. "Liquid Adsorption Chromatography in Columns and on Thin Layers," *J. Chromatog.*, *38*, 209 (1968).
3. Snyder, L. R. "Column Efficiencies in Liquid Adsorption Chromatography: Past, Present and Future," *J. Chromatogr. Sci.*, *7*, 352 (1969).
4. Majors, R. E. "High Performance Liquid Chromatography on Small Particle Silica Gel," *Anal. Chem.*, *44*, 1722 (1972).
5. Kirkland, J. J. "High Performance Liquid Chromatography with Porous Silica Microspheres," *J. Chromatogr. Sci.*, *10*, 593 (1972).
6. Scott, C. D. *Clin. Chem.*, *14*, 521 (1968).
7. Huber, J. F. K. "Theory and Practice of Column-Liquid Chromatography," *Chimia*, Suppl., 24 (1970).
8. Majors, R. E. "Effect of Particle Size on Column Efficiency in Liquid-Solid Chromatography," *J. Chromatogr. Sci.*, *11*, 88 (1973).
9. Majors, R. E. and F. R. MacDonald. "Practical Implications of Modern Liquid Chromatographic Column Performance," *J. Chromatog.*, *83*, 169 (1973).
10. Halasz, I. and I. Sebastian. "New Stationary Phase for Chromatography," *Angew. Chem., Int. Ed. Engl.*, *8*, 453 (1969).

11. Aue, W. A. and C. R. Hastings. "Preparation and Chromatographic Uses of Surface-Bonded Silicones," *J. Chromatog.*, *42*, 319 (1969).
12. Kirkland, J. J. and J. J. DeStefano. "Controlled Surface Porosity Supports with Chemically-Bonded Organic Stationary Phases for Gas and Liquid Chromatography," *J. Chromatogr. Sci.*, *8*, 309 (1970).
13. Kirkland, J. J. "High Speed Liquid-Partition Chromatography with Chemically Bonded Organic Stationary Phases," *J. Chromatogr. Sci.*, *9*, 206 (1971).
14. Bossart, C. J. "Chemically Bonded Liquid-Phase Packings for Gas Chromatography," *ISA Transactions*, *7*, 283 (1968).
15. Locke, D. C., J. T. Schmermund, and B. Banner. "Bonded Stationary Phases for Chromatography," *Anal. Chem.*, *44*, 90 (1972).
16. Majors, R. E. "Techniques for Liquid Chromatographic Columns Packed with Small Porous Particles," *Anal. Chem.*, *45*, 755 (1973).
17. Majors, R. E. "High Performance Liquid Chromatography on Small Particle Alumina," submitted for publication in *J. Chromatogr. Sci.*
18. Huber, J. F. K. "Mass Transport and Distribution in the Chromatographic Process," presented at the Discussions of the German Bunsen Society, Saarbrucken, W. Germany, October, 1972.
19. Locke, D. C. "Chemically Bonded Stationary Phases for Liquid Chromatography," *J. Chromatogr. Sci.*, *11*, 120 (1973).
20. Schmit, J., R. A. Henry, R. C. Williams, and J. F. Dieckman. "Applications of High Speed Reversed-Phase Liquid Chromatography," *J. Chromatogr. Sci.*, *9*, 645 (1971).
21. Knox, J. H. and G. Vasvari. "The Performance of Packings in High Speed Liquid Chromatography. III. Chemically Bonded Pellicular Materials," *J. Chromatog.*, *83*, 181 (1973).
22. Novotny, M., S. L. Bektesh, K. B. Denson, K. Grohmann, and W. Parr. "Polar Silicone-Based Chemically Bonded Stationary Phases for Liquid Chromatography," *Anal. Chem.*, *45*, 971 (1973).
23. Novotny, M., S. L. Bektesh, and K. Grohmann. "Chemically Bonded Stationary Phases with Variable Selectivity," *J. Chromatog.*, *83*, 25 (1973).
24. Snyder, L. R. *Principles of Adsorption Chromatography*. (New York: Marcel Dekker, Inc., 1968).
25. "Spectroquality Solvents Chart," published by MCB Manufacturing Chemists, Norwood, Ohio, 1972.
26. Majors, R. E. "High Speed Liquid Chromatography of Antioxidants and Plasticizers Using Solid Core Supports," *J. Chromatogr., Sci.*, *8*, 338 (1970).
27. Burtis, C. A. and D. R. Gere. *Nucleic Acid Constituents by Liquid Chromatography*. (Walnut Creek, California: Varian Aerograph, 1970).

CHAPTER 9

USE OF NOVEL SILANES FOR THE SOLID-PHASE PEPTIDE SYNTHESIS AND THE PREPARATION OF POLAR CHEMICALLY BONDED PHASES FOR LIQUID CHROMATOGRAPHY

Wolfgang Parr and Milos Novotny

INTRODUCTION

The synthetic possibilities for peptides and other macromolecules of biological importance have been tremendously advanced by the methods developed by Merrifield.[1-6] The basic idea of a specific and relatively stable attachment of suitable reagents to a solid matrix followed by a build-up of peptide chains of more or less desirable length through the successive coupling of individual amino acids has overcome cumbersome fragment condensations encountered in the classical peptide synthesis. While this method has not yet achieved a full degree of maturity due to primarily technological reasons, it has been successfully applied to many difficult synthetic problems, and its potential for the synthesis of larger macromolecules as well as the tremendous acceleration of the synthetic process in general is now obvious. The most common matrices used in the Merrifield solid-phase synthesis are polymeric beads of the styrene-divinylbenzene type[7] which are subsequently chloromethylated by the Friedel-Crafts reaction or otherwise modified.

Naturally, the physical properties of polymeric beads are quite critical for their synthetic function. Solubility factors, swelling, extent of cross-linking, and concentration of chlorine atoms throughout the bead may significantly influence the results of synthesis. Thus, it is not surprising that the list of different polymer materials hitherto examined for solid-phase synthesis is quite long.[6] The most severe problems encountered with such polymer beads are those associated with the processes of diffusion and mass transfer of solvents and reagents into and out of the polymer structure. The problems with the distribution of growing peptide chains throughout the whole bead matrix were pointed out by Merrifield and Littau[8] as based on the autoradiography measurements.

Bayer *et al.*[9] reported that a possible formation of failure sequences with the Merrifield method has its origin in the limited mass transfer into the polymeric beads, also. As these failure sequence problems increase with the growing chain of the synthesized polypeptide, the development of new suitable materials is needed for achieving further progress in this area.

Several approaches hitherto suggested which may lead to improved results in the peptide synthesis include the use of rigid copolymers with larger pores and specific areas,[10,11] macroreticular resins,[12] resin-coated glass beads,[13-16] and controlled-thickness organic materials permanently bonded to inorganic matrices.[16-20]

In the present stage of development, many common features are shared between solid-phase synthesis and modern liquid chromatography. Likewise, sorption-desorption kinetics, resistance to mass transfer, the diffusion of chromatographed solutes into and out of the porous materials have been of major concern for improving the resolving power of chromatographic columns. The efforts to develop macroporous adsorbents for chromatography,[21,22] use of smaller particles,[23-25] advances in ion-exchange chromatographic materials,[26] pellicular packings,[13,27] and chemically-bonded phases[16,28-32] quite closely parallel the general trend of improved peptide synthesis.

While only future developments will decide their relative importance in these areas, certain directions for achieving these goals have already been indicated. The analogy between the two fields has been, perhaps, most clearly shown in the work by Bayer *et al.*,[9] and recently by Scott *et al.*,[14,15] with the pellicular resins originally designed[13] for high-pressure liquid chromatography. Their work even suggests that chromatography be used as an effective method for evaluation of the materials and their suitability for the Merrifield synthesis. The properties of resin-coated glass beads, which are important in peptide synthesis, were investigated.[9,14] It was shown that under optimized conditions, the resistance to mass transfer can be decreased by one to two orders of magnitude with the pellicular resins, thus resulting in the production of several peptides of higher purity in considerably shorter times. The higher permeability of the pellicular materials further permits the synthesis to take place in a column and allows automation of the whole synthetic process.[15] The approach of obtaining more favorable mass-transfer rates with smaller particle sizes, now so popular in high-resolution liquid chromatography,[23-25] is not applicable to Merrifield's resins because of permeability reasons.

Another promising approach to the solution of mass-transfer problems in the solid-phase peptide synthesis exists in the utilization of suitable reactive organic layers permanently bonded to inorganic matrices such as porous silica, glass beads, and porous glass. Once again, this approach received attention through advances in the technology of chromatographic columns.[16] The reactivity of surface silanol groups of such materials permits a number of surface modifications. Esterification,[16] Grignard substitution,[33] and silylation[28-32,34,35] have been thus far the most commonly explored procedures for the preparation of chemically-bonded phases for chromatography. As a function of the reagents chosen and the reaction conditions, the surface structures can be envisioned as either monomolecular layers or films of polymeric materials of different consistency and more or less controlled thickness. In any event, the reactive groups are now located and ready for direct interaction on the surface or in close vicinity within the polymeric layer. Furthermore, such materials may be superior in their performance to the resins mechanically coated on the surface of glass beads.

Following the chromatographic work of Halász and Sebastian,[16] Bayer *et al.*[17] chose an esterification of a porous silica material with 1,4-dihydroxymethyl benzene for the synthesis of the model peptides $(Leu\text{-}Ala)_6$ and (Leu-Leu-Glu-Gly). All reaction times were reduced in comparison with the "classical" beads and no failure sequences were detected. However, the limited hydrolytical stability of esterified siliceous surfaces is a serious drawback of this attachment method. This difficulty can be overcome by bonding organic molecules to the silica materials through Si-O-Si-C bonds which are more stable against an attack by electrophilic or nucleophilic agents.

While there is essentially no limit in the selection of proper reagents to form esterified layers, silane compounds possessing desirable functional groups in their side chain are not commonly met. Therefore, Parr and Grohmann[18-20] synthesized model silane compounds which comply with the requirements of the solid-phase synthesis and provide further a hydrolytically stable linkage between the siliceous inorganic matrices and the bonded organic layer. The syntheses of these novel silanes, their chemical and physical properties, and the reactions with surface structures will be described in this article together with the selected examples of peptides synthesized through these methods.

Since the synthetic methods described herein can provide the surface structures

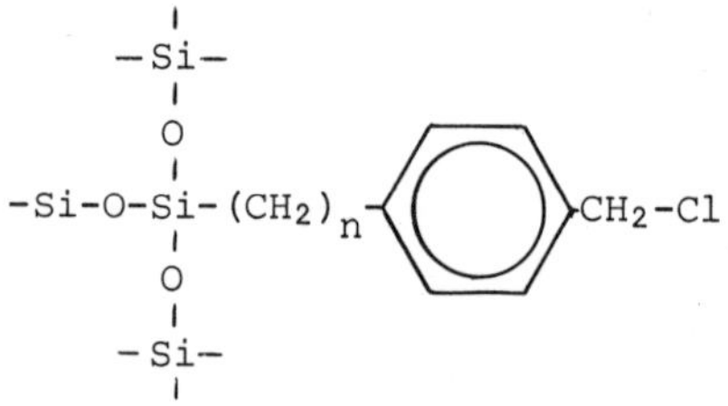

$$\begin{array}{c} | \\ -Si- \\ | \\ O \\ | \\ -Si-O-Si-(CH_2)_n-C_6H_4-CH_2-Cl \\ | \\ O \\ | \\ -Si- \\ | \end{array}$$

it is obvious that the terminal chlorine can be modified by simple chemical reactions to create surface films of different polarities. Thus, we have once again arrived at the situation where the technology of peptide synthesis and that of chromatographic columns complement each other. Developments in the area of chromatographic columns for both gas and liquid chromatography during the last several years had indeed concentrated to a great degree on the preparation of chemically bonded stationary phases. While in numerous cases such bonded column substrates have not quite lived up to expectations of alleviating numerous practical problems, many interesting packings with unusual properties have been obtained. In particular, many practical advantages were achieved with the reversed-phase systems using the packings with chemically bonded hydrophobic films. However, relatively little had been reported on the preparation of polar silicone bonded phases,[28,30,36] thus reflecting the limited number of suitable silane compounds readily available. The novel silanes synthesized by Parr and Grohmann[16,19,20] give almost universal possibilities for the preparation of the polar chemically bonded phases of different selectivities. Novotny *et al.*[31,32] have prepared several packings with different selectivities for liquid chromatography and studied their properties. This work will be reviewed in the present chapter.

SYNTHESIS OF NOVEL SILANES

There are several methods available for the formation of siloxane bonds with the surface silanol groups of glass.[17,28,32-35] However, only halosilanes have the tendency to bind spontaneously to the surface at room temperature. This is one of the reasons why substituted halosilanes have been selected in this study.

For peptide synthesis, as well as for chromatographic supports, substituted mono-, di-, and trichlorosilanes are of interest since this flexibility in the functional groups permits the formation of both monomolecular layers and highly crosslinked structures. The novel silanes can be derived from the following general formula:

$$Y'-\underset{Y''}{\overset{Y}{Si}}-(CH_2)_n-C_6H_4-CH_2-X$$

n = 0,1,2.... X = Cl or Br Y' = Y" = Cl or CH_3
Y = Cl

The general approach to the synthesis of these silanes is the following: synthesis of the proper alkyl side chain first, and attachment to the chlorosilane last. Since it is very difficult to purify chlorosilane derivatives because of the hydrolytic properties, this approach guarantees a product of high purity. Additionally, purification other than distillation cannot be carried out. At present, there are only two methods available for the preparation of substituted halosilanes on a laboratory scale. In the first approach a suitably substituted Grignard reagent is reacted with tetrachlorosilane.[37] In this reaction impurities of the type — R_2SiCl_2, R_3SiCl, and R_4Si — are always present. Fractional distillation has to be used to purify the intermediates as well as the final product.

Nevertheless, we have used this approach to synthesize 4-bromomethylphenyltrichlorosilane. p-Tolyl-magnesium bromide was reacted with silicon tetrachloride according to the method of Chvalovsky *et al.*[38] The obtained 4-tolyl-trichlorosilane was then brominated with N-bromosuccinimide to yield the desired product. After fractional distillation, 4-bromomethylphenyltrichlorosilane was obtained in 52% yield (B.P. 125° C/2mm Hg). The NMR spectrum showed the expected signals (δ = 4.42 ppm, S, 2H, and δ = 7.43-7.75 ppm, Q, 4H) and a reasonable purity of the monomer.

The second method calls for catalytical addition of halosilanes to alkenes. Catalysts for this reaction are *tert*-amines or hexachloroplatinic acid.[37,39] This technique has the advantage that the desired olefin can be added to the selected halosilanes and removal of impurities by distillation assures a product of high purity.

In order to minimize complications during the solid-phase synthesis the benzyl ester linkage has been selected for the growing peptide chain, since the behavior of this group has been very well established.

The desired olefins can be prepared according to the following scheme:

$$X-C_6H_4-(CH_2)_n-X \xrightarrow[\text{Allylbromide}]{\text{Mg/Ether}} X-C_6H_4-(CH_2)_n-CH\ -CH=CH_2$$

$$\xrightarrow[\text{2) } CH_2O(g)]{\text{1) Mg/Ether}} HO-CH_2-C_6H_4-(CH_2)_n-CH_2-CH=CH \xrightarrow{\text{HCl/anhydrous } Na_2SO_4}$$

$$Cl-CH_2-C_6H_4-(CH_2)_n-CH_2-CH=CH_2 + H-Si(Y)(Y')(Y'') \xrightarrow{H_2PtCl_6}$$

$$Cl-CH_2-C_6H_4-(CH_2)_n-CH_2-CH_2-CH_2-Si(Y)(Y')(Y'')$$

7 6 5 4 3 2 1

n = 0,1,2,3..... Y = Cl Y' = Y" = Cl or CH_3

Various halosilanes have been prepared according to this scheme:

[I] trichloro-[3-(4-chloromethylphenyl)propyl]silane

n = 0 Y = Y' = Y" = Cl

[II] methyldichloro-[4-(4-chloromethylphenyl)butyl]silane

n = 1 Y = Y' = Cl Y" = CH_3

[III] dimethylchloro[4-(4-chloromethylphenyl)butyl]silane

n = 1 Y = Cl Y' = Y" = CH_3

1,4-Dibromobenzene and 4-chlorobenzylchloride are converted by a Wurtz synthesis with allylbromide and magnesium into 1-allyl-4-bromo-benzene and 4-(4-chlorophenyl-1-butene), respectively, which can be transformed via a second Grignard reaction with formaldehyde(g) into the corresponding hydroxy compounds (1-allyl-4-hydroxymethylbenzene) and 4-(4-hydroxymethylphenyl)-1-butene, respectively. Conversion of these hydroxy compounds into the corresponding chloro derivatives can be achieved with hydrogen chloride(g) in the presence of anhydrous Na_2SO_4 (6 hours, room temperature). Catalytical addition in the presence of hexachloroplatinic acid of these olefins to trichlorosilane leads to the formation of [I]. Reaction of 4-(4-chloromethylphenyl)-1-butene with methyldichlorosilane or dimethylchlorosilane yields [II] or [III], respectively. The yields in the individual steps and the boiling points of the respective compounds are listed in Table 9.1.

All intermediates as well as the end products have been characterized by means of NMR-spectroscopy. The assignment of the chemical shifts of the individual protons has been made by comparison of the known spectra of allylbenzene,4-phenyl-1-butene, benzylchloride,and the different substituted chlorosilanes. Integration of the signals showed the correct ratio of the protons. As can be seen from Table 9.2, no addition of HCl to the carbon-carbon double bond takes place during the nucleophilic substitution of the hydroxy group. Furthermore, the data indicate that no isomerization of the double bond can be found. In addition, the NMR spectra confirmed that the catalytic addition of silanes to the substituted propene and butene yielded only the terminal isomer. For further identification, the final products have been investigated by gas chromatography and mass spectrometry. An LKB 9000 gas chromatograph-mass

Table 9.1

Physical Properties of Synthetic Intermediates and Final Products

Compound	*Yield*	*B.P., °C*
1-Allyl-4-bromo-benzene	73%	51°C/0.01 mm Hg (lit 105°C/15mm Hg)[40]
4-(4-Chlorophenyl)-1-butene	63%	53°C/0.05 Hg (lit 93°C/10mm Hg)[41]
1-Allyl-4-hydroxymethylbenzene	40%	98°-103°C/0.75 mm (lit 80°C/0.01 mm Hg)[40]
4-(4-Hydroxymethylphenyl)-1-butene	65%	85°C/0.25 mm Hg
[I]	48%	125°C/0.05 mm Hg
[II]	58%	122-125°C/0.8 mm Hg
[III]	63%	123°C/0.4 mm Hg

spectrometer was used to carry out the experiments. The compounds were chromatographed on a glass column (1m x 4mm) on Chromosorb P(80-100 mesh) coated with 1% OV-1. All three monomers (I-III) showed impurities of less than one per cent. A typical mass spectrum is shown in Figure 9.1. The spectrum of dichloromethyl[4-4-chloromethylphenyl]butylsilane is representative for all compounds prepared by this method. The molecular ion is found at m/e 294. The peaks at m/e 295, 296, 297, 298, 299, 300, and 301 correspond to the isotopes for ^{13}C, ^{2}H, ^{29}Si, ^{30}Si, and ^{37}Cl. Other ion fragments show also a similar isotopic distribution. Therefore, this technique can be used as a direct structure proof. Details concerning the structure elucidation utilizing the ratio of these isotopes have been described.[20] Table 9.3 shows a formal interpretation of the mass spectrum in Figure 9.1 and, again, is representative for all other compounds described in this study.

Table 9.2

^{1}H-Parameter of the Compounds I, II, and III
[60 MHz, δppm, TMs = 0]

		*1**	*2*	*3*	*4*	*5*	*6*	*7*
I	n = 0 Y=Y'=Y"=Cl	---	1.20-1.30 (2H,M)	1.80-1.90 (2H,M)	2.60-2.70 (2H,M)		7.15-7.30 (4H,Q)	4.5 (2H,S)
II	n = 1 Y'Y'=Cl Y"=CH_3	0.7 (3H,S)	1.15-1.25 (2H,M)	1.65-1.90 (4H,M)		2.60-2.80 (2H,M)	7.15-7.30 (4H,Q)	4.50 (2H,S)
III	n = 1 Y=Cl Y'=Y"=CH_3	0.35 (6H,S)	1.1-1.2 (2H,M)	1.55-1.75 (4H,M)		2.60-2.75 (2H,M)	7.10-7.25 (4H,Q)	4.45 (2H,S)

*The numbering of the protons corresponds to those in the reaction scheme.

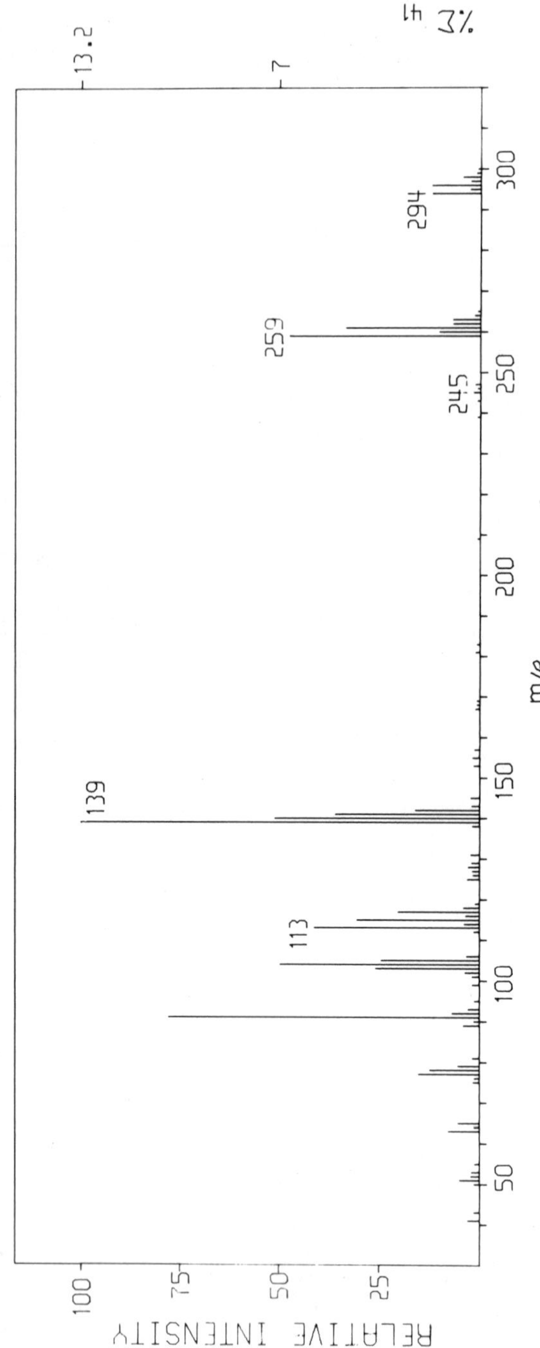

Figure 9.1. Mass spectrum of Dichloromethyl-[4-chloromethylphenyl)butyl]silane at 70 eV. Instrument LKB 9000; Column : (1m x 4mm) glass; Packed with Chromosorb P (80-100 mesh) Coated with 1% OV-1; Temperature programmed, 5%/min, starting at 130°C; Injector, 200°C; Separator, 225°C; Ion source, 270°C; Accelerating voltage, 3.5 kV.

Table 9.3

Formal Interpretation of the Mass Spectrum of Methyldichloro-[4-(4-chloromethylphenyl)butyl] silane

Cl	CH_2	C_6H_4	CH_2	CH_2	CH_2	CH_2	$Si(Cl)_2CH_3$
35	49	125	139	151	165	181	294
294	259	245	169	155	141	127	113

m/e	ion/fragment
294	M
259	M-Cl
245	$M-CH_2-Cl$
181	$M-Si(Cl)_2-CH_3$
139	$Cl-CH_2-C_6H_4-CH_2$
113	$SiCl_2-CH_3$

PREPARATION OF CHEMICALLY BONDED PHASES AND THEIR PROPERTIES

There are several types of siliceous materials which can be utilized for the preparation of chromatographic packings with bonded layers of stationary phases. While the bonded polar silicone polymers as synthesized with the present silanes may also find some use in gas-chromatographic applications, our experience has been limited so far to liquid chromatography, and porous and superfically-porous spherical particles. Porasil C (37-75 μ) and Corasil I (37-50 μ) have been used throughout most of our experiments, but the silanes described in this article can also be reacted with small-particle silica gel (5-10 μ) for obtaining better chromatographic efficiencies.

Depending on the type of siliceous material, 5-10% w/w of the silanes in dry tetrahydrofuran or benzene is reacted with the supports for a required period of time. The details of the bonding procedure have been described.[31] The reaction times needed for a complete reaction may vary with the type of selected silane (trichloro-, dichloro-, monochloro-derivative, or their respective mixtures), and this step still remains to be optimized. Reactivity of these silanes is predominantly influenced by the degree of substitution on the silicon atom, and the number of methylene units between the silicon atom and the aromatic ring. Chlorine analyses obtained by the Volhard titration method indicated that the reaction of surface silanol groups of Porasil C with trichloro[4-(4-chloromethylphenyl)butyl] silane or their mixtures was completed by approximately 50% after 12 hours at 40°C. Even longer reaction times would be necessary with such "bulky" molecules for a more complete reaction. The unreacted and unbonded material is extracted from the beads exhaustively with a number of solvents of different polarity.

In order to prepare stationary phases with different selectivities, simple organic reactions are utilized for the treatment of $-C_6H_4-CH_2-Cl$ residues in the bonded layers. Since many of these polymeric layers are of a resinous nature, a 1:1 mixture of water and tetrahydrofuran has been most commonly used for good wettability and penetration of reagents into the polymeric structures. During various modifications of the terminal groups, one must bear in mind also the stability of Si-O-Si-C links with the basic silica framework. It has been our experience that these bonded films are stable at a pH as low as 1, but the depolymerization process is likely at the extremes of alkalinity. However, slightly alkaline solutions allow many reactions to proceed more or less quantitatively without any appreciable damage to -Si-O-Si-C bonds. Thus, for instance, a hydrolysis with a slightly alkaline solution of sodium carbonate in a mixture of water and tetrahydrofuran at 40°C leads regularly to the formation of a hydroxy phase[31] with good reproducibility.[32] This reaction seems to proceed quite rapidly as evidenced by the evolution of gas and a change in affinity of the beads to the hydrolytic solution. After the selective modification is completed, the materials are successively refluxed and washed with a series of solvents of different polarities and dried.

Different ratios of silane monomers will primarily dictate the final consistency and physical properties of the resultant polymeric layers as they are decisive for the degree of formation of linear versus three-dimensional polymers, the degree of cross-linking, chain terminations, etc. The effects of monomer blending on the chromatographic properties of nonpolar bonded silicones were studied by Aue and Hastings.[29] Optimum conditions obviously must be found empirically with any given set of silane monomers, their proportions, type of solid support, and reaction conditions. Changes in the column selectivity of the hydroxy phase due to different proportions of monomers are reflected in Table 9.4 where the capacity ratio values, k, were measured with model compounds of different polarity. Porasil C alone (an adsorbent of about 50 m^2/g) is included for comparative purposes. While these results caution that the factors influencing the conditions of polymerization must be brought under strict control, and possibly optimized in any given case, the potential utilization of such phenomena to control the column selectivity is strongly suggested.

Table 9.4

k-Values for Columns Prepared by Different Procedures

Column No.	Packing	k-Values: o-Cresol	k-Values: Phenol	k-Values: Benzyl Alcohol	k-Values: Phenanthrene-quinone-
1	Porasil C	3.06	6.04	9.85	11.70
2	50% dichloro- and 50% trichloromonomer; not polymerized at high temperature	2.90	14.00	12.70	10.40
3	20% trichloromonomer and 80% dimethyl-dichloro-silane	4.15	9.20	12.00	12.60
4	100% dichloromonomer	4.73	13.70	10.70	11.60
5	100% trichloromonomer	5.20	14.00	14.30	15.10
6	As 2, but polymerized at high temperature	4.40	10.20	12.20	13.30

A 1:1 ratio of trichloro[3-(4-chloromethylphenyl) propyl] silane and dichloromethyl[4-(4-chloromethylphenyl)butyl] silane has been chosen by ourselves for further syntheses of cyano and ester phases.[32] Cyano phases were prepared by the treatment of bonded $-C_6H_4-CH_2-Cl$ residues with potassium cyanide dissolved in tetrahydrofuran and water. To obtain an ester phase, the hydroxylated polymeric film was further subjected to the reaction with benzoyl chloride in tetrahydrofuran in the presence of a small amount of pyridine as a catalyst. Table 9.5 gives the comparison of the column selectivities for nitroaniline isomers, thus reflecting a degree of specific interaction (mostly hydrogen bonding). As can be seen from Table 9.5, the degree of selectivity is quite predictable in these cases. A distinct advantage of the novel silanes described in this chapter is that a number of selective modifications can essentially be made from the same material already bonded to siliceous matrices. Thus, the variables due to different reactivities of various starting monomers and their degree of polymerization are eliminated. This is best reflected in the reproducibility of the preparation of cyano and hydroxy phases as indicated in Table 9.5. The modifications involving multiple-step reactions result in less reproducibility. A number of other modifications to prepare various selective packings can be made as based on this approach (for instance, amines, amides, or ion-exchangers can be prepared in a similar fashion). Even more attractive is the preparation of chromatographic materials with extremes of selectivity or even specificity as suggested by the recent work of Grushka and Scott.[42] Biologically important molecules could be synthesized or suitably attached on such silylated surfaces in order to study specific interactions between peptides and proteins, polynucleotides, between drugs or hormones and their receptors, etc. Due to the column permeability reasons, inorganic matrices may in the future seriously challenge gel materials used predominantly in these areas of research. Permeability of such materials may also be the key to easier automation of the relevant procedures in large columns, thus scaling up the purification methods for biological molecules into the large preparative range.

Just as with the pellicular resins,[14] the thicknesses of the interacting polymeric layers must be optimized for both peptide synthesis and chromatography. Our investigations[31,32] as well as the previous work done with the chemically bonded silicones[30,43] suggest

Table 9.5

Relative Selectivities of the Columns with Bonded Films of Different Polarities

Monomer	*Modification of the Stationary Phase*	*Functional Groups*	*k-Value*		
			o-Nitroaniline	*m-Nitroaniline*	*p-Nitroaniline*
None*	None	Surface silanol groups	0.95	2.74	7.70
Trichloro-[3-(4-chloromethylphenyl) propyl]silane and dichloromethyl[4-(4-chloromethylphenyl)-butyl]silane(1:1)	Hydrolysis with a slightly alkaline solution	Hydroxy	0.95	2.96	8.45
Same	Reaction with potassium cyanide	Cyano	1.16 1.14 1.13	3.08 3.06 3.18	9.50 9.50 9.55
Same	Hydrolysis and subsequent esterification of the hydroxy phase	Ester	1.04	2.71	8.40
Monochlorodimethyl-[4-(4-chloromethyl-phenyl)butyl]silane	Reaction with potassium cyanide	Cyano	1.11	3.30	9.07

*Unmodified Porasil used for comparative purposes.

that we are dealing with resinous materials rather than liquids, and a certain degree of solvation with a polar component ("moderator") is essential for the satisfactory chromatographic performance. It can be seen from Figure 9.2 that the number of theoretical plates per second increases with the percentage of a "moderator," and that this increase is more dramatic with the stationary phase which is expected to be more penetrable for the solvent. On the other hand, it is well known that the difference in polarity of the two phases in the classical liquid-liquid chromatography forms the basis of an effective separation. Consequently, a loss of selectivity is to be expected due to addition of a polar mobile phase constituent in any practical work with the polar chemically bonded stationary phases. A decrease of k-values

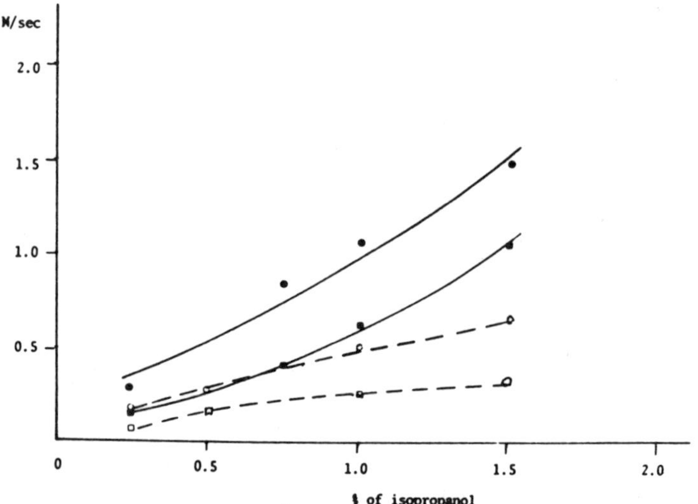

Figure 9.2. *Dependence of the number of theoretical plates per second on concentration of isopropanol in the mobile phase.*
Conditions: Columns (2mm, i.d. x 80cm) with different hydroxypolymers bonded on Porasil C (37-75 μ); Flow rate 1.0 ml/min; Temperature 23°C. (———) column with the ratio of dichloro- and trichloromonomer 1:1, not polymerized at high temperature. (------) column with the ratio of dichloro- and trichloromonomer 1:1, polymerized at high temperature
(O) Phenol (□) Cresol.
[Reproduced from Anal. Chem., 45, 971 (1973) through the permission of the American Chemical Society.]

with the amount of isopropanol added to the mobile phase (n-heptane) for phenol and cresol on the hydroxy phase is shown in Figure 9.3. This effect is again different for the columns with expected different consistency of the polymeric layer.

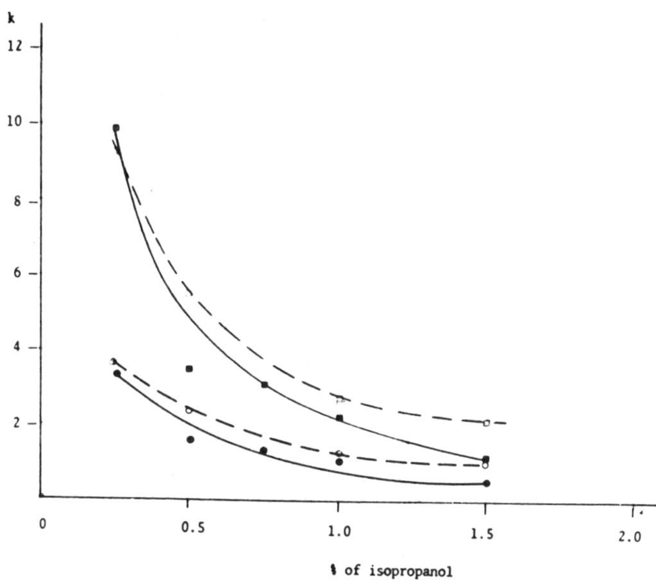

Figure 9.3. *Dependence of k-values on concentration of isopropanol. Conditions and symbols as in Figure 9.2. [Reproduced from Anal. Chem., 45, 971 (1973) through the permission of the American Chemical Society.]*

Obviously, a compromise between the column efficiency and selectivity must be chosen depending on a particular application. Selectivity of the separation process with bonded phases can also be influenced by temperature.[31,44] Preliminary results[32] suggest that monomolecular layers as prepared through the reaction of siliceous surfaces with monochlorodimethyl[4-(4-chloromethylphenyl)butyl] silane followed by a suitable modification may reduce some of these problems generally encountered with chemically bonded stationary phases in both gas and liquid chromatography.

SYNTHESIS OF PEPTIDES

Several different types of porous silica beads can be used for the preparation of solid supports for peptide synthesis. Porasil E (800-1500 Å) and porous glass beads (Bioglass 2500 Å) have been used throughout our experiments. All support material is reacted with 5% solution of the monomer in benzene. Details of the polymerization procedures have been described,[18-20] but they have not been optimized. In principle, it is true that these supports can be prepared on the same basis as the chemically bonded phases for liquid chromatography. The capacity of the beads is in the range of 0.03-0.15 meq/g of Cl or Br as determined by Volhard analysis. Table 9.6 shows the range of the obtained values for the coatings with the individual monomers.

Table 9.6

Halogen Content of Silanized Glass Beads

	(Bioglass 2500 or Porasil E)	*meq Br or Cl/g*
A	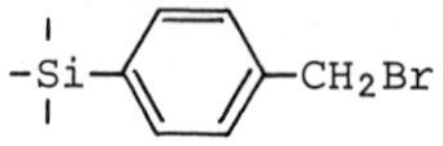 $-Si-C_6H_4-CH_2Br$	0.08-0.15
B	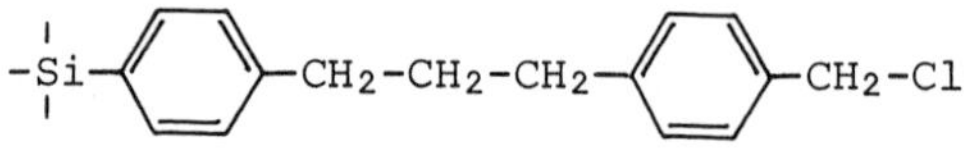 $-Si-C_6H_4-CH_2-CH_2-CH_2-C_6H_4-CH_2-Cl$	0.03-0.08
C	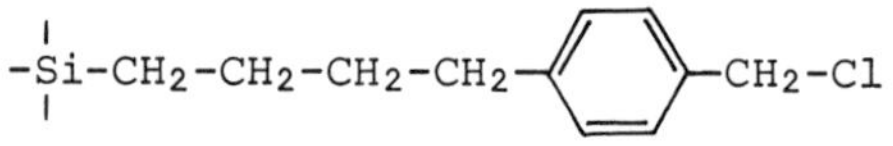 $-Si-CH_2-CH_2-CH_2-CH_2-C_6H_4-CH_2-Cl$	0.03-0.08
D	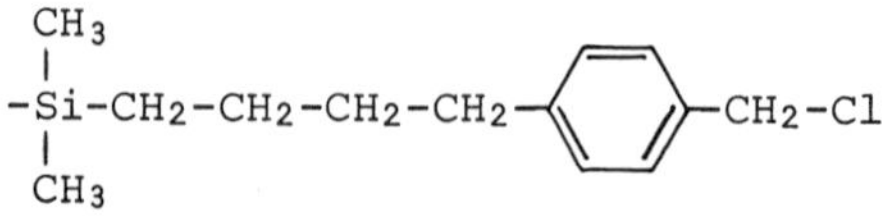 $-Si(CH_3)_2-CH_2-CH_2-CH_2-CH_2-C_6H_4-CH_2-Cl$	0.03-0.07

In order to demonstrate the suitability of these supports for the solid phase synthesis, three different peptides have been prepared utilizing the supports A and B. These three peptides should provide answers to the following questions:

1. Is a synthesis on this type of support possible?
2. Are failure sequences due to minimized diffusion problems reduced?
3. Can the generally applied batch procedure be replaced by a column procedure analog to a set-up as in liquid chromatography?

The general outline of the solid phase synthesis is shown in Figure 9.4 and is followed throughout these experiments.

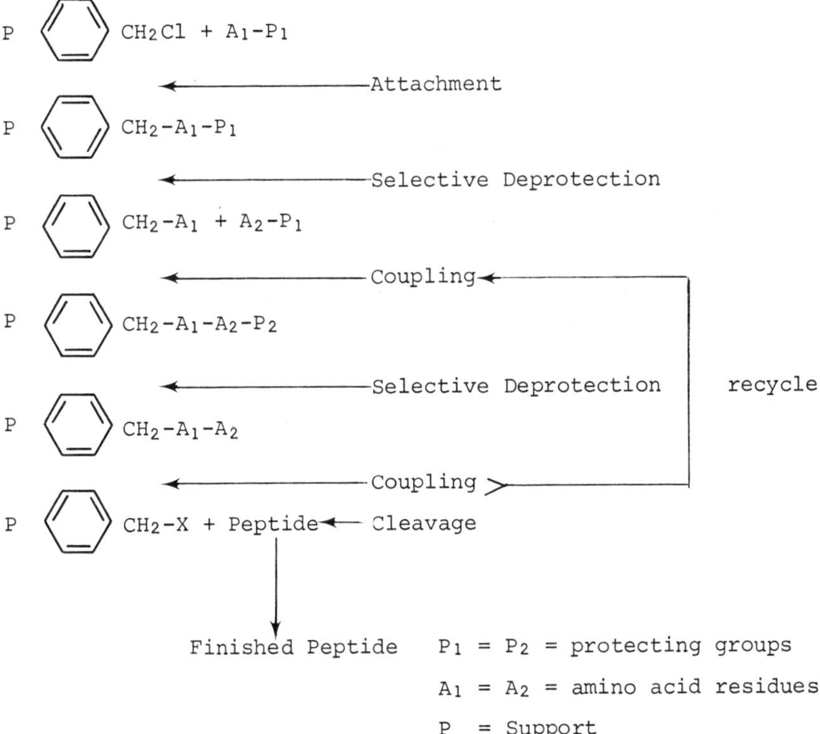

Figure 9.4. General outline of the solid phase peptide synthesis.

The first model peptide synthesized has been the tetrapeptide H-Pro-Gly-Phe-Ala-OH. All steps have been carried out as previously described for the general procedures of solid phase peptide synthesis. Two stability tests have been applied in order to show the hydrolytic resistance of these materials.

After the attachment of the first amino acid, the amino acyl support was treated for 24 hours with trifluoroacetic acid and subsequently for 24 hours with triethylamine/$CHCl_3$. Both reagents are essential for cleavage and neutralization reactions during the synthesis of peptides. Thin-layer chromatography has shown that no free amino acid is liberated during the treatment, thus proving the stability of the Si-O-Si-C bonds.

After cleavage of the synthesized tetrapeptide, which was purified and identified by thin-layer chromatography and amino acid analysis, it has been found that the cleavage did not proceed to completion (Table 9.7). Total hydrolysis of peptide remaining on the carrier showed only 50% conversion. These difficulties can be explained by the negative effect of electron withdrawing of the SiO_3 group in the para-position to the benzyl ester linkage. These difficulties can be overcome if a short aliphatic chain is inserted between the three-dimensional silicate network and the benzyl group as in compounds B, C, and D.

As a model peptide using support B, we synthesized the dodecapeptide H-(Phe-Ala)$_6$-OH, which at the same time can be used to indicate failure sequences formed due to incomplete coupling reactions. Failure sequences are easily detected since in a dodecapeptide with a repeating sequence the failure sequences are multiplied. The synthesis was carried out as outlined earlier and details have been described.[19] Cleavage of the dodecapeptide from the carrier was, as expected, practically complete (~90%). The crude product was partially hydrolyzed, esterified and trifluoroacetylated. Through the combination gas chromatography-mass spectrometry, no dipeptides of the type H-Ala-Ala-OH or H-Phe-Phe-OH were observed, indicating that no failure sequences were present. One has to consider these results as semi-quantitative since partial hydrolysis does not yield only dipeptides. In addition, free amino acids are formed and also longer peptides which cannot be detected by gas chromatography due to low volatility.

In order to find an answer to our third question, we tested the suitability of the silicone support B for column procedures. All synthetic steps beginning with the attachment of the second amino acid were carried out in a column procedure as outlined in Figure 9.5. The peptide to be synthesized possessed the primary structure H-Gln-Gln-Gly-Gly-Tyr-NH_2.

Table 9.7

Summary of the Model Peptides Synthesized Using Novel Silicon Matrices

Glass Beads		*Total Capacity*	*Total Capacity of Attachment of First Amino Acid*	*Peptide Synthesized*	*Coupling Technique*	*Yield*	*Identification of Final Products*
Type	*Capacity*						
A	0.13mMol Br/g	5g=0.65mMol	0.43mMol Ala	H-Pro-Gly-Phe-Ala-OH	DCC[a]	100mg(50%)	TLC,AAA[c]
B	0.075mMol Cl/g	8.9g=0.67mMol	0.24mMol Ala	H-(Phe-Ala)$_6$OH	DCC	240mg(~90%)	TLC,AAA,GC-MS
C	0.082mMol Cl/g	30g-2.5mMol	1.2mMol Tyr	H-Gln-Gln-Gly-Gly-TyrNH_2	ONP[b]	600mg(90%)	TLC,AAA,GC-MS

[a]DCC = Dicyclohexylcarbodiimide

[b]ONP = p-Nitrophenyl ester

[c]AAA = Amino acid analysis

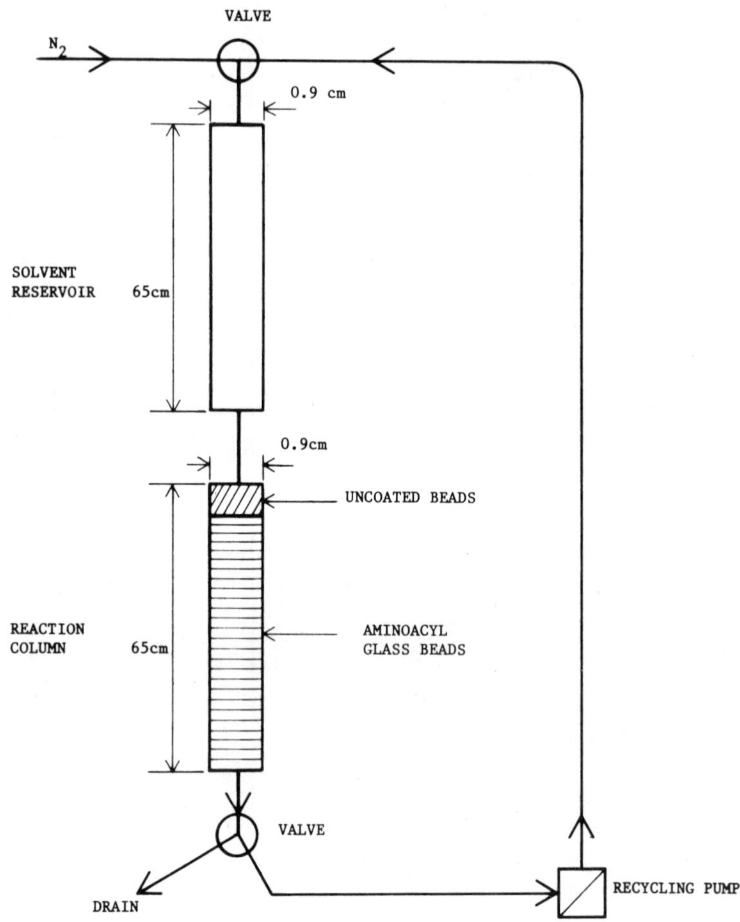

Figure 9.5. Apparatus for the synthesis of peptides with silicon matrices.

This peptide was selected since an authentic sample from the classical fragment condensation was available for comparison.

A chromatographic column (65 cm x 0.9 cm) was filled with the amino acyl beads. Dead volume was eliminated by placing uncoated beads on top of the filling. A second column of the same size was used as a reservoir (Figure 9.5). All operations except for the individual coupling reactions were carried out by forcing N_2 through the column. The effluent on the drain side can be collected and checked for completeness of washing or coupling steps. In order to prevent build-up of dicyclohexylurea, p-nitrophenyl ester coupling reactions were used exclusively (Table 9.7).

The advantage of a p-nitrophenyl ester coupling is that the p-nitrophenol liberated during the reaction can be monitored in order to test for completeness of the individual coupling steps. Monitoring can be carried out easily just by taking aliquot samples on the drain side. The measured absorbance at 315 nm is a direct indication of the progress of coupling. In all our experiments the measurements indicated that the reactions were completed after approximately 6 hours.

The peptide was cleaved from the solid support by transesterification with methanol-trimethylamine for 24 hours; again monitoring at 280 nm indicated that the transesterification was complete after 24 hours. The crude product was purified by gel permeation chromatography. After amminolysis in a mixture of DMF-methanol and cleavage of the protecting groups by catalytic hydrogenation, the unprotected peptide was again purified by gel permeation chromatography. The peptide eluted as a single peak and proved to be homogeneous in various solvent systems and undistinguishable from an authentic sample.

Grushka and Scott[42] recently used a commercially available silane, 1-trimethoxysilyl-2(4-chloromethylphenyl)ethane (from Union Carbide), for the preparation of a new type of permanently bonded phase suitable for peptide synthesis. However, these authors were not interested in the preparation of the peptide (*i.e.*, purity, absence of failure sequences). The peptides were left attached to the support and were used as stationary phases for liquid chromatography.

The presented results show that solid phase peptide synthesis can be carried out on the siliceous matrices with functional groups on the surface. However, longer peptides have to be synthesized and more sensitive analytical methods developed before the advantages and disadvantages of these inorganic matrices can be judged.

ACKNOWLEDGMENT

This work was supported by National Science Foundation Grants No. GP-33751 (to M. Novotny) and GP-26019 (to W. Parr). The support of the Robert A. Welch Foundation (Grant E-404), Houston, Texas, is also acknowledged.

REFERENCES

1. Merrifield, R. B. "Solid Phase Peptide Synthesis. I. The Synthesis of a Tetrapeptide," *J. Amer. Chem. Soc.*, *85*, 2149 (1963).
2. Merrifield, R. B. "Solid Phase Peptide Synthesis. II. The Synthesis of Bradykinin," *J. Amer. Chem. Soc.*, *86*, 304 (1964).
3. Merrifield, R. B. "Solid Phase Synthesis. III. An Improved Synthesis of Bradykinin," *Biochemistry*, *3*, 1385 (1964).
4. Merrifield, R. B. "Automatic Synthesis of Peptides," *Science*, *150*, 178 (1965).
5. Merrifield, R. B. "Solid Phase Peptide Synthesis," *Adv. Enzymol.*, *32*, 221 (1969).
6. Marshall, G. R. and R. B. Merrifield. in *Biochemical Aspects of Reactions on Solid Supports*, G. R. Stark (Ed.), (New York and London: Academic Press, 1971), p. 111.
7. Merrifield, R. B. "Peptide Synthesis on a Solid Polymer," *Proc., Fed. Amer. Soc. Exp. Biol.*, *21*, 412 (1962).
8. Merrifield, R. B. and V. Littau. "Solid Phase Peptide Synthesis. The Distribution of Peptide Chains on the Solid Support," E. Bricas (Ed.), (1968), p. 179.
9. Bayer, E., H. Eckstein, K. Haegele, W. A. Koenig, W. Bruening, H. Hagenmeier, and W. Parr. "Failure Sequences in the Solid Phase Synthesis of Polypeptides," *J. Amer. Chem. Soc.*, *92*, 1738 (1970).
10. Merrifield, R. B. "Solid Phase Peptide Synthesis," *Endeavour*, *24*, 3 (1965).
11. Millar, J. R., D. G. Smith, W. E. Marr, and T. R. E. Kressman. "Solvent Modified Polymer Network," *J. Chem. Soc. (London)*, *218*, (1963).
12. Kunin, R., E. Meitzner, and N. Bortnick. "Macroreticular Ion Exchange Resins. Part I. The Preparation and Characterization of Expanded Network and Macroporous Styrene Divinylbenzene Copolymer and their Sulphonate," *J. Amer. Chem. Soc.*, *84*, 305 (1962).
13. Horvath, C. G., B. A. Preiss, and S. R. Lipsky. "Fast Liquid Chromatography: An Investigation of Operating Parameter and the Separation of Nucleotides by Ion Exchange," *Anal. Chem.*, *39*, 1422 (1967).
14. Scott, R. P. W., Ka Kong Chan, P. Kucera, and S. Zolty. "The Use of Resin Coated Glass Beads in the Form of Packed Bed for the Solid Phase Synthesis of Peptide," *J. Chromatogr. Sci.*, *9*, 577 (1971).
15. Scott, R. P. W., S. Zolty, and Ka Kong Chan. "An Automatic Apparatus for the Synthesis of Peptides Using Resin Coated Glass Beads in the Form of a Packed Bed," *J. Chromatogr. Sci.*, *10*, 384 (1972).
16. Parr, W. and K. Grohmann. "Festphasen-Peptidsynthese an einer anorganischen Matrix mit organischen gruppen an

der Oberfläche," *Angew. Chem. Int. Ed. Engl.*, *11*, 314 (1972).

17. Halász, I. and I. Sebastian. "New Stationary Phases for Chromatography," *Angew. Chem. Int. Ed. Engl.*, *8*, 453 (1969).
18. Bayer, E., G. Jung, I. Halász, and I. Sebastian. "A New Support for Polypeptide Synthesis in Columns," *Tetrahedron Lett.*, *4503* (1970).
19. Parr, W. and K. Grohmann. "A New Solid Support for Peptide Synthesis," *Tetrahedron Lett.*, *28*, 2633 (1971).
20. Parr, W., K. Grohmann, and K. Haegele. "Silicon-Matrizen für die Festphasen-Peptidsynthese," *Ann. Chem.*, in press.
21. Kiselev, A. V. "Adsorbents in Gas Chromatography," *Adv. Chromatogr.*, *4*, 113 (1967).
22. Bebris, N. K., A. V. Kiselev, B. Y. Mokeev, Y. S. Nikitin, Y. I. Yashin, and G. E. Zaizeva. "Macroporous Silica as an Adsorbent for Molecular Chromatography," *Chromatographia*, *4*, 93 (1971).
23. Huber, J. F. K. "Theorie und Praxis der Säulen Flüssigkeit Chromatographie," *Chimia, Suppl.*, *24*, (1970).
24. Majors, R. E. "High Performance Liquid Chromatography on Small Particle Silica Gel," *Anal. Chem.*, *44*, 1722 (1972).
25. Kirkland, J. J. "High Performance Liquid Chromatography with Porous Silica Microspheres," *J. Chromatogr. Sci.*, *10*, 593 (1972).
26. Benson, J. V. and J. A. Patterson. in *New Techniques in Amino Acids, Peptide and Protein Analysis*, A. Niederwieser and G. Pataki (Eds.) (Ann Arbor, Mich: Ann Arbor Science Publishers, 1971).
27. Kirkland, J. J. "Controlled Surface Porosity Support for High Speed G. L. C.," *Anal. Chem.*, *41*, 218 (1969).
28. Bossart, C. J. "Chemically Bonded Liquid Phase Packing for Gas Chromatography," *ISA Trans.*, *7*, 283 (1968).
29. Aue, W. A. and C. R. Hastings. "Preparation and Chromatographic Uses of Surface-Bonded Silicon," *J. Chromatogr.*, *42*, 319 (1969).
30. Kirkland, J. J. and J. J. DeStefano. "Controlled Surface Porosity with Chemically Bonded Organic Stationary Phases for Gas and Liquid Chromatography," *J. Chromatogr. Sci.*, *8*, 309 (1970).
31. Novotny, M., S. L. Bektesh, K. B. Denson, K. Grohmann, and W. Parr. "Polar Silicone-Based Chemically Bonded Stationary Phases for Liquid Chromatography," *Anal. Chem.*, *45*, 971 (1973).
32. Novotny, M., S. L. Bektesh, and K. Grohmann. "Chemically Bonded Stationary Phases with Variable Selectivity," *J. Chromatogr.*, *83*, 25 (1973).

33. Locke, D. C., J. J. Schermund, and B. Banner. "Bonded Stationary Phases for Chromatography," *Anal. Chem., 44,* 90 (1972).
34. Abel, E. W., Z. H. Pollard, P. C. Uden, and G. J. Nickless. "A New Gas Liquid Chromatographic Phase," *J. Chromatogr., 22,* 23 (1966).
35. Stewart, H. N. M. and S. G. Perry. "A New Approach to Liquid Partition Chromatography," *J. Chromatogr., 37,* 97 (1968).
36. Majors, R. E. and J. J. Hopper. 160th National Meeting, American Chemical Society, Chicago, Illinois, September 1970, paper no. A42.
37. Bazant, V. and V. Chvalovsky. *Chemistry of Organosilicon Compounds.* (New York: Academic Press, Vol. I, 1965). Ch. Advances in A.A. and Peptide Analysis Using Spherical Resins and the Application in Biochemistry and Medicine, p. 180.
38. Chvalovsky, V. and V. Bazant. "Organosilicon Compounds. Fisson of Aryl-Silicon Bonds by Means of Nitric Acid," *Coll. Czechosl. Chem. Comm., 16,* 580 (1951).
39. Noll, W. *Chemistry and Technology of Silicones.* (New York: Academic Press, 1968). Chapter 8.-"Properties of Technical Products, pp. 446-447; Chapter 10. "Silicones in the Glass Industry and Ceramics," pp. 579-585.
40. Belg. Pat. 660565, National Research and Development Corporation (1965).
41. Martin, M. M. and G. J. Glercher. "The Addition of Trichloromethyl Radical to Substitute 3 Phenyl-1-propenes and 4-phenyl-1-butenes," *J. Amer. Chem. Sac., 86,* 233 (1964).
42. Grushka, E. and R. P. W. Scott. "Polypeptide as a Permanently Bound Stationary Phase in Liquid Chromatography," *Anal. Chem., 45,* 1626 (1973).
43. Kirkland, J. J. "High Speed Liquid Partition Chromatography with Chemically Bonded Organic Stationary Phases," *J. Chromatogr. Sci., 9,* 206 (1971).
44. Schmit, J. A., R. A. Henry, R. C. Williams, and J. F. Dieckman. "Application of High Speed Reversed-Phase Liquid Chromatography," *J. Chromatogr. Sci., 9,* 645 (1971).

CHAPTER 10

CHROMATOGRAPHY ON SILYLATED SURFACES

Milos Novotny

INTRODUCTION

Various siliceous materials have been used in chemical separations since the very early days of chromatography. With advancing technology, other materials of chromatographic interest have also become increasingly available. Nevertheless, even today, siliceous surfaces of different natures vastly dominate nearly all areas of gas and liquid chromatography. The applications of such materials to other fields (such as heterogeneous catalysis and chemical engineering processes) complement their chromatographic use, and chromatography can be useful even in model studies of relevant phenomena.[1] Increasing use of physicochemical methods in surface studies, in general, has just begun to extend our understanding of the nature of siliceous materials, their inner structure, pore-size distribution and the chemistry of the surface structures.

In terms of chromatographic utility, the siliceous materials have been employed primarily as solid supports for partition chromatography (GLC and LLC) and adsorbents (LSC and GSC). While naturally occurring materials possessing a basic silica framework as well as artificially prepared adsorbents found their extensive use even in their unmodified forms, practical chromatographers are constantly aware of the necessity for certain modifications in order to prepare suitable materials for analytical purposes. In addition, many physical chemists who study sorbate-sorbent interactions in general have experienced a great need for well-defined and "tailor-made" surfaces.

Fundamental studies of these aspects were initiated largely due to the efforts of Kiselev and his coworkers (for a review see references 2 and 3) and many important directions pointed out based on their work. In addition to geometrical modifications of such surfaces to suit a given separation problem, the chemical treatment of surface structures was stressed[4] in order to cope with many fundamental problems encountered in the practical chromatography. While the reactivity of surface silanol groups allows several chemical modifications, various silylation treatments became a natural choice due to the stability of Si-O-Si-C linkages.

While unnoticed by many chromatographers, Kiselev's earlier systematic studies and pointed directions have since been paralleled by ever increasing efforts to improve present technology of chromatographic columns in terms of efficiency, selectivity and inertness toward the chromatographed solutes. Preparation of inert solid supports for gas-liquid chromatography, advances in modern liquid chromatography and the recent interest in glass capillary columns for the separation of biological compounds once again focused considerable attention on the silylation of different siliceous surfaces. While the utilization of a stable silane link between silica matrices and the rest of the bonded organophilic layer may be of fundamental scientific and technological importance in a number of cases ranging from improved glass wettability[5] to the synthesis of macromolecules,[6] we will limit this discussion to the three areas which appear most pertinent to the problem of chemical separations: (a) deactivation of solid supports, (b) preparation of chemically bonded stationary phases and modified adsorbents, and (c) surface wettability as connected with the performance of chromatographic columns.

BASIC CHEMISTRY OF SILICEOUS SURFACES

Although new spectroscopic analytical methods for surface studies have just started to penetrate the areas pertinent to chromatography, and our understanding of siliceous surfaces and their reactivity is far from complete, much useful information has been accumulated through more conventional measurements and empirical observations of many analysts. The materials most commonly used in chromatography are silica gels, porous glasses, diatomaceous earths, ceramic and glass beads, and glass capillary columns. Although the common feature of all of them is the basic silica framework, their surface chemistry is often distinctly different, depending on surface geometry, pore-size distribution, additional elements, and their position within the silica structure, and type and amount of surface-active groups. All these properties contribute not only to different chromatographic behavior of such materials, but also to different types of both reversible and irreversible interactions with various chemical compounds.

In addition to naturally-occurring silicates which must undergo a series of treatments before usable for chromatography, silica gels can be prepared synthetically from either water glass or $SiCl_4$. Surface heterogeneity is usually a characteristic feature of common silica gels which must be reduced by various treatments to suit a given chromatographic problem. Thus, while unmodified silica gels are suitable for the separation of light molecules, their small pores present great difficulties for larger or polar molecules in both gas and liquid chromatography. Therefore, considerable attention has been paid to the preparation of macroporous materials through the thermal treatments[2,4,7,8] which produce larger pores and may reduce the values of specific surface from several hundred m^2/g down to 25 m^2/g or even less. Even more flexibility in the geometrical modification procedures is given by special glass materials (*e.g.*, sodium borosilicate glass) which can be leached and heat-treated to provide chromatographic packings with rather easily controlled structures. Surfaces with narrow pore-size ranges can now be prepared with the pore diameters up to several thousand angströms. The initial composition of glass, the extent of heat treatment, and etching exert primary control over surface character and chromatographic performance. Tremendous utility of porous glass in the area of macromolecular fractionations is now well-established, and its potential for faster large-scale operations is quite obvious due to its rigidity and small resistance to flow. In a similar fashion, both ordinary and porous glass capillary tubes can be modified for either adsorption[4,9-11] or partition chromatography.[11,12]

Surface chemical structures of siliceous materials are of primary importance in chromatography. Until more information on these subjects becomes available, certain assumptions concerning the arrangements and types of surface silica structures must be made. The spacings between SiO_2 tetrahedra and their regularity are expected to be different in the case of pure silica and other materials. For instance, in case of different glasses the distribution of alkaline and other oxides within the silicon-oxygen network is a rather important factor. For the sake of simplication, a planar model with standard bond distances may be considered:

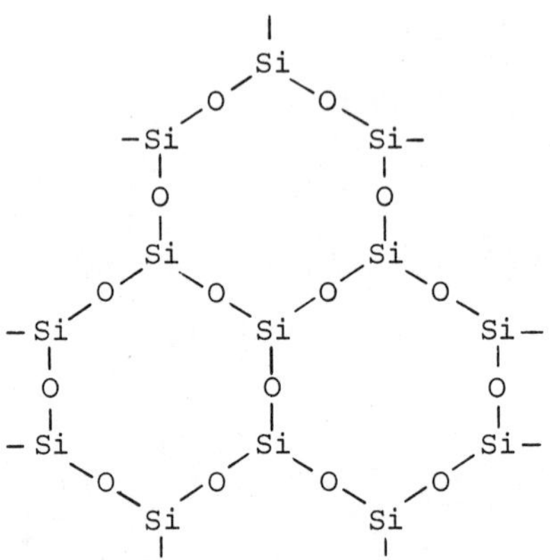

While this structure, when fully hydroxylated, can theoretically accommodate eight hydroxy groups per $100 Å^2$,[13] actual numbers as reported by various authors and based on different experimental techniques[14-17] are smaller. Different types of surface hydroxylated structures are likely[18] in addition to a certain number of hydrogen-free -Si-O-Si (with O bridge) bridges.

According to Kiselev,[2] surface silanol groups are primarily responsible for the specific solute-sorbent interactions with silica gel, and wide-pore materials with decreased surface area obtained through hydrothermal treatment[19] lose their character of a specific adsorbent. Specific adsorption due to the presence of silanol groups has been proved spectrally.[20] The surface silanol groups exhibit quite strong acidic effects[21] due to their protonation and can be looked upon as weak ion exchangers. According to Iler,[13] the equilibrium constant for $-\text{Si-OH} \rightleftarrows -\text{Si-O}^{-}+\text{H}^{+}$ is of the order of 10^{-10} — an acid strength comparable to phenol.

It is now established that silanol groups and other acidic sites of silica surfaces are primarily responsible for the adsorption phenomena which are often undesirable in chromatographic separations. Due to their presence, an analyst too often faces the problem of sample losses and unreliable quantitation, tailing peaks, catalyzed sample conversions, decreased lifetime of liquid stationary phases, etc. There are several options for the blocking of silanol groups by a suitable chemical reaction such as an esterification, Grignard reaction, or silanization. Thus far, silanization appears to provide the most realistic approach to blocking of surface silanol groups due to the hydrolytical and thermal stability of Si-O-Si-C linkages. These bonds are readily ruptured only in strong alkaline media, and they are quite stable even at relatively low pH values. Silylated surfaces have been shown to be stable up to 350°C.[22] A variety of silyl compounds with different degrees of substitution have been shown to be applicable to different forms of chromatography since the first use of dimethyldichlorosilane by Howard and Martin[23] for the preparation of a reversed-phase packing for liquid chromatography of C_{12}-C_{18} fatty acids. In addition to the character of initial siliceous surface, the course of silylation is dependent primarily on the type of silane and reaction conditions. It can be generally stated that the reactivity decreases in the following order:

trichlorosilane > dichlorosilane >

monochlorosilane > alkoxysilanes.

Furthermore, the length of side chain and the type of substitution influence significantly the reaction rates. Various degrees of surface conversion have been observed with different silanes.[24-26] Differently substituted silanes can be used to comply with the purpose of surface modification. Thus, for instance, while trisubstituted chlorosilanes should essentially provide only a monomolecular surface coverage, the formation of polymeric layers will take place with di- and trichlorosilanes. This appears to be a desirable feature of these reactions for creating a variety of chromatographic packings for different separation purposes. On the other hand, more insight into the structures of these substrates is increasingly needed to provide us with a better understanding of the solute-substrate interactions.

The three main areas of the utilization of silylation reactions in chromatography are: deactivation of solid supports, preparation of chemically bonded stationary phases (or modified adsorbents), and the preparation of "tailor-made" siliceous surfaces for improved wettability. Chronologically, this is also the order in which the emphasis on these developments has gradually developed, and they will be treated so in this chapter.

DEACTIVATION OF SOLID SUPPORTS

With the early emphasis on petrochemical applications of gas-liquid chromatography and only gradual increase of interest in the separation of more polar and heavier molecules, it is not surprising that the problems of deactivation of solid supports did not become urgent until the beginning of the 1960's. Diatomaceous earths had been initially chosen as solid supports for gas-liquid chromatography because of their relatively low surface area. Due to their biologic origin, diatomaceous earth materials exhibit a large degree of heteorogeneity in both their morphology and surface chemistry. According to Ottenstein,[7] some 10,000 diatoms may have contributed to the formation of materials now in daily use in many laboratories. Besides silica, which creates approximately 90% of the initial material, various other oxides are present which influence the inertness of supports in practical applications and make the deactivation task difficult. The content of ingredients may vary as dependent on a given resource, manufacturing process, and further treatment. The surface area is significantly reduced by the thermal treatment, and many undesirable ingredients are removed by further alkaline and/or acid leaching processes. It is desirable that the solid support has a relatively small specific surface area (with no more than several m^2/g) and small pores are totally absent. Much useful information on the surface geometry of diatomaceous supports became available through scanning electron microscopy.[27-29]

Liquid-phase distribution on the solid support is very critical for the chromatographic performance. As pointed out by Giddings,[30] most chromatographic theories do not assume the real distribution of stationary liquids on diatomaceous supports, but rather operate with their average thickness. According to Giddings' theoretical model, the total liquid coated on a solid support can be divided into the portion adsorbed on the overall surface and the capillary liquid condensed in the pores. The extent of this distribution does not depend only on the geometrical character of the solid surface, but also on the nature of surface structures which may interact to a different degree with the coated liquid. The problems of compatibility of solid surface and the coated liquid which primarily touch upon the wettability phenomena are common to capillary columns and solid supports. They will be discussed in the latter part of this chapter.

Although the deactivation problems can often be solved for individual types of analyses by different means,[7] a satisfactory general approach to this problem has not arrived as yet. The heterogeneity of the diatomaceous surface makes the task of deactivation particularly difficult. The kinetic data measured by Perrett and Purnell,[31] suggest that more than one type of adsorptive site may be present on the surface of such materials. When approaching the task of surface deactivation with thus far available analytical surface methodology, gross simplification must be made which may not always be applicable.

Although one must be aware of the possible existence of different surface groups,[3,18] for the sake of simplification any hydroxylated siliceous material may be envisioned as:

```
 OH   OH   OH   OH   OH
 |    |    |    |    |
-Si-O-Si-O-Si-O-Si-O-Si-O-
 |    |    |    |    |
 O    O    O    O    O
 |    |    |    |    |
```

The use of surface silanization for deactivation purposes had arisen from the practical problems of tailing and sample decomposition[32] which are particularly critical with lightly-loaded columns. Several silanization reagents and many empirical procedures have been proposed since. The general silylation reaction between the hydrolylated silica surface and a chlorosilane

```
             R                      R
 |           |      -HCl            |
-Si-OH + Cl-Si-R ----------> -Si-O-Si-R
 |           |                      |
             R                      R
```

leads theoretically to the surface coverage with an organic monolayer. Completeness of this reaction depends on the type of silyl donor and reaction conditions. On the model of fully hydroxylated surface, it is not hard to see that complete coverage cannot be expected for steric reasons even with trimethylchlorosilane. "Bulky" phenylsilanes must, therefore, provide considerably less substitution,[24] and also the chain length and the presence of functional groups in the side chain will have a definite influence on the reactivity of silane compounds. On the other hand, such compounds when bonded to only some silicon atoms of the surface may still quite effectively hinder any bigger solute molecules from the silica framework, thus providing an efficient deactivation. Methylsilanes have been used thus far exclusively for the deactivation purposes. However, the silanes possessing selective groups are worthwhile to explore (better deactivation is likely to depend on better surface wettability with a given stationary phase).

Hexamethyldisilazane[33-35] and dimethyldichlorosilane[32,36,37] have been chosen as alternatives to monochlorosilanes, as it is expected that they provide more "dense" monolayers due to their easier reaction with vicinal surface hydroxy groups:

```
                  CH3    CH3
  |    |           |      |          -NH3        |            |
-Si-O-Si- +CH3-Si-NH-Si-CH3 ---------> -Si — O — Si-
  |    |           |      |                      |            |
  OH   OH         CH3    CH3                     O            O
                                                 |            |
                                        CH3-Si-CH3 CH3-Si-CH3
                                                 |            |
                                                CH3          CH3
```

```
                     Cl
  |    |             |         -2HCl         |            |
-Si-O-Si- + Cl-Si-CH3 ----------->          Si — O — Si-
  |    |             |                       |            |
  OH   OH           CH3                      O            O
                                              \          /
                                                  Si
                                              /          \
                                            CH3          CH3
```

Borisenko, *et al.*[26] compared the extent of chemical modification of silica gel with trimethylchlorosilane, hexamethyldisilazane and dimethyldichlorosilane. These authors found the latter two compounds to be more effective than trimethylchlorosilane which was not sufficient to remove all surface silanol groups. It has been a common experience that dimethyldichlorosilane is the most efficient silylation agent used in the surface deactivation of diatomaceous earths

materials, and this reaction has been used extensively by the manufacturers.[7]

The reaction with dimethyldichlorosilane is of particular interest. While used commonly by both manufacturers and researchers, its mechanism cannot be fully explained on the basis of the earlier quoted reaction. As suggested already by Bohemen, *et al.*,[33] the reactive chlorines may be present in the surface layer after the treatment because two vicinal groups may not always be available for interaction:

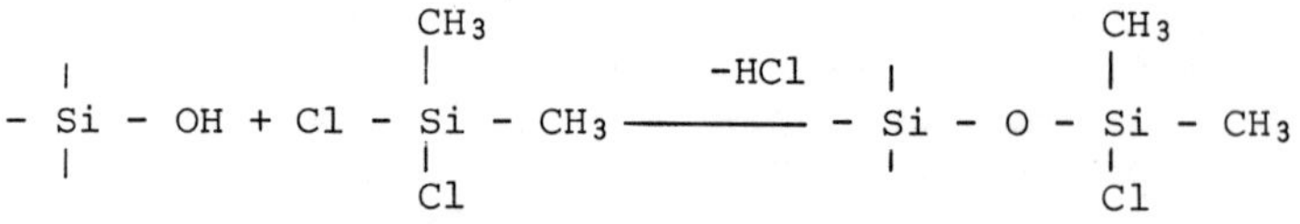

$$
\begin{array}{c}
 \\ | \\ -\,Si\,-\,OH \\ | \\
\end{array}
+
\begin{array}{c}
CH_3 \\ | \\ Cl\,-\,Si\,-\,CH_3 \\ | \\ Cl
\end{array}
\xrightarrow{-HCl}
\begin{array}{c}
CH_3 \\ \qquad\qquad | \\ -\,Si\,-\,O\,-\,Si\,-\,CH_3 \\ \qquad\qquad | \\ Cl
\end{array}
$$

Washing with an alcohol, which removes HCl from the surface and, furthermore, hydrolyzes the remaining chlorine residues, was suggested.[36] While this hypothesis would seem quite plausible, the recent paper by Gilpin and Burke[17] appears to challenge this idea. According to the results of these authors, based on careful measurements, the following mechanism seems to operate during the secondary treatment with an alcohol:

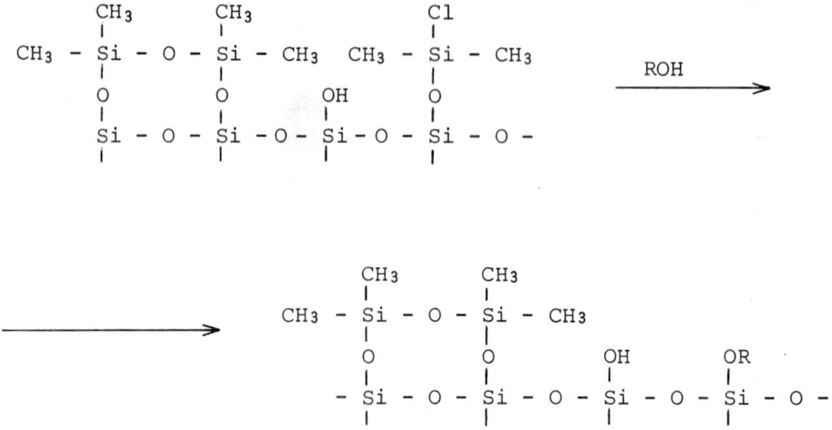

$$
\begin{array}{ccccccccccccccc}
 & & CH_3 & & & & CH_3 & & & & & & Cl & & \\
 & & | & & & & | & & & & & & | & & \\
CH_3 & - & Si & - & O & - & Si & - & CH_3 & & CH_3 & - & Si & - & CH_3 \\
 & & | & & & & | & & & & & & | & & \\
 & & O & & & & O & & OH & & & & O & & \\
 & & | & & & & | & & | & & & & | & & \\
 & & Si & - & O & - & Si & - O - & Si & - & O & - & Si & - & O\,- \\
 & & | & & & & | & & | & & & & | & &
\end{array}
\xrightarrow{ROH}
$$

$$
\longrightarrow
\begin{array}{cccccccccccccccc}
 & & CH_3 & & & & CH_3 & & & & & & & & & \\
 & & | & & & & | & & & & & & & & & \\
CH_3 & - & Si & - & O & - & Si & - & CH_3 & & & & & & & \\
 & & | & & & & | & & & & & & & & & \\
 & & O & & & & O & & & & OH & & & & OR & \\
 & & | & & & & | & & & & | & & & & | & \\
 & - & Si & - & O & - & Si & - & O & - & Si & - & O & - & Si & - O - \\
 & & | & & & & | & & & & | & & & & | &
\end{array}
$$

It would practically mean that the surface treated in this fashion is not as inert and stable as commonly believed.

In addition to the mechanism discussed above, it should be born in mind that the traces of water

present on the siliceous surfaces will induce the formation of a linear polymer linked to the surface. This mechanism which has been suggested[2] is consistent with the observation of Holmes and Stack[36] that higher concentrations of dimethyldichlorosilane used for the surface treatment prior to coating resulted in tailing of sterols.

It can be stated that the present stage of technology of deactivated supports is sufficient for many types of stationary phases and applications. However, with increasing demands for more sensitive detection, further developments in this area will be needed in the near future. Consequently, a search for better defined siliceous surfaces may be expected together with efforts directed to a better understanding of the surface phenomena.

CHEMICALLY BONDED STATIONARY PHASES

In case of the deactivation of solid supports for chromatography we have been primarily concerned with blocking of silanol groups and, presumably, the formation of monolayers. However, when reactive intermediates are attached to the siliceous surface, further reactions can take place which may lead either to an increased molecular weight of the monolayer residues or the polymerization process. In the latter case, bulky layers of about 1μ can be created[29,38] which are still quite firmly attached to the silica matrix and practically nonextractable with solvents of different polarities. The control of surface attachment and the following reactions of the intermediates with other molecules, modifications, polymerizations, etc., may be important keys to the solution of fundamental processes desirable not only for chromatographic purposes, but also for the controlled synthesis of macromolecules,[6,39,40] preparative-scale purifications of biologic polymers, catalysis, ion-exchange processes, wettability problems,[5,41,42] etc. Therefore, it is highly desirable to obtain sufficient knowledge of various chemical events taking place on the surface and to master the synthetic art leading to a desired product.

In spite of increased activity in this field during the last several years, the separation mechanism in many such chromatographic media is not yet fully understood. A sharp distinction between a "modified adsorbent" and a "chemically bonded

```
        R              R
        |              |
Cl - Si - Cl   Cl - Si - Cl                      R               R
        |              |                         |               |
        Cl             Cl                   - O - Si --- O ---- Si - O -
               +                                 |               |
        Cl             Cl          +H2O          O               O
        |              |          ------>        |               |
CH3- Si - CH3  CH3- Si - CH3              CH3- Si - CH3  CH3- Si - CH3
        |              |                         |               |
        O              O                         O               O
        |              |                         |               |
     - Si ---- O ---- Si -                    - Si --- O ---- Si -
        |              |                         |               |
```

and the remaining hydroxy groups deactivated in a usual way. The synthesized materials were essentially nonextractable and performed well in the chromatography of numerous compounds (hydrocarbons, alcohols, derivatized amino acids and nucleic bases, pesticides, etc.). For example, Figure 10.1 shows a chromatogram of N-trifluoroacetyl, n-butyl esters of nineteen amino acids.

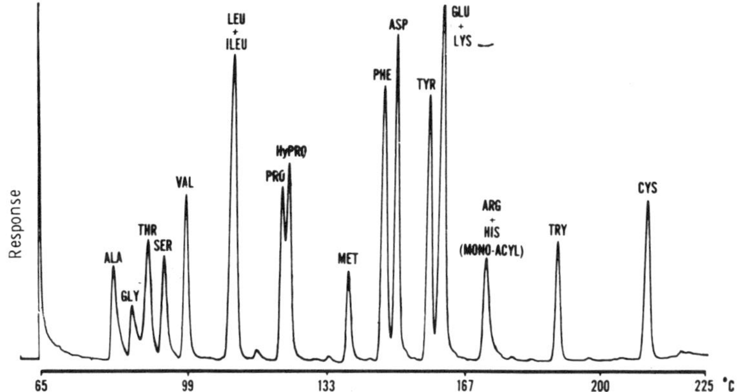

Fiqure 10.1. Chromatograph of N-trifluoroacetyl, n-butyl esters of 19 amino acids.
Column: 13.2% w/w of $(C_{18}H_{37}SiO_{3/2})n$ on 60-80 mesh Chromosorb G.
Column: 1m x 4mm, i.d., (glass)
Programming rate: 6°C/min
Nitrogen flow rate: 45 ml/min
The injection mixture contained about 5 μg of each amino acid.
[Reproduced from Aue, W. A. and C. R. Hastings, J. Chromatogr., 42, 319 (1969) with permission of Elsevier Publishing Company.]

Polymerizations with both long-[38] and short-chain[44] chlorosilanes have been described which yielded satisfactory chromatographic materials. The problems with high volatility of short-chain silanes were overcome by partial hydrolysis prior to polymerization.[44] Hastings *et al.*[45] have further demonstrated that the whole synthetic procedure can be carried out in already pre-packed columns, but the fluidized-bed method applied earlier[38,44] gave superior results. It was shown that the bonding capability is dependent on the chain length of reacting silanes and the polymerization conditions.[38,44] However, it is necessary to stress that optimum conditions should be sought for any given set of starting materials. While quite extensive studies were performed with the nonpolar bonded phases, much less is known about polar bonded substrates and their chromatographic properties. Bossart[47] tested several commercially available alkoxysilanes containing polar groups with variable success and noticed their limited stability. Kirkland and DeStefano[49] prepared "ether" and "nitrile" phases bonded to the controlled-surface porosity supports for the primary use in liquid chromatography, but showed also some gas-chromatographic applications. It can be noticed that relatively high temperatures are needed to bring about the elution of common substances. Just as with other materials of somewhat similar properties,[54] the polar silicone-based phases seem to be more effective for the separations of low-boiling compounds.[49,55] The performance of silicone bonded phases in gas chromatography is dependent primarily on their physical consistency and selectivity. Both properties are strongly dependent on the nature of original silanes, degree of substitution, functional groups and length chain, extent of polymerization and degree of cross-linking. Many ordinary silicone polymers (SE-30, SE-52, OV-1, etc.) appear as gummy materials at the room temperature where their chromatographic performance is impaired due to the high viscosity and resulting resistance to mass transfer. It is similarly expected that the bonded phases should possess limited viscosity at the column operating temperatures for good chromatographic results.

Aue *et al.*[38,44] found that some column efficiencies for different solutes and van Deemter curves were not greatly inferior to those obtained with coated silicone elastomers. The bonded layers obtained from prehydrolyzed low-molecular-weight silanes[44] actually resembled the properties of SE-30 (a silicone gum with approximately expected structural similarities). On the other hand, bonded phases may possess the qualities not available with ordinary silicones, such as unusual types of selectivity,[46] and temperature stability[44] which may make them suitable for high-sensitivity detection and gas chromatography-mass spectrometry. Detailed investigations are necessary to understand the solute-substrate interactions encountered with the bonded phases and to be able to control effectively the polymer formation on a support particle. Aue's work[29,38] suggests that the polymeric layers prepared in his laboratory are of the order of 10^{-6} meter thick and their molecular weight is of the order of at least 10^4. This estimate is supported by the data obtained by the elemental analysis, thermogravimetry, and scanning electron microscopy.[29,44,45] Furthermore, gel permeation chromatographic measurements have shown the extractable portions of these polymers to be between 10^3-10^5 of molecular weight.[45] The distribution of bonded nonpolar silicones on the surface of diatomaceous supports is particularly interesting. The data suggest that, unlike in the conventional gas-liquid chromatography,[29] the bonded polymer is primarily placed on the macrosurface (periphery) of the solid particles and, if present in the pores altogether, its thickness must be small. Good chromatographic performance combined with this fact suggests that the mass-transfer phenomena occur only in the surface layers and the chromatographed molecules do not penetrate into the whole depth of the stationary phase. The polymeric layers bonded to the solid support form impressive views (Figure 10.2) which "are somewhat reminiscent of a rocky landscape covered by a heavy snowfall."[29]

Liquid Chromatography

The necessity of permanently modifying the siliceous support in liquid chromatography for some applications had been realized as early as 1950 by Howard and Martin.[23] However, even now, the vast majority of practical separations are carried out with liquid-solid chromatography. The classical adsorbents often possess a unique selectivity toward the chromatographed compounds, but unfortunately of only quite limited range. The necessary adjustments of the adsorbent activity (*e.g.*, through the addition of water) or the mobile-phase composition are often irreproducible and troublesome. Moreover, the extremes of selectivity cannot be achieved in this way.

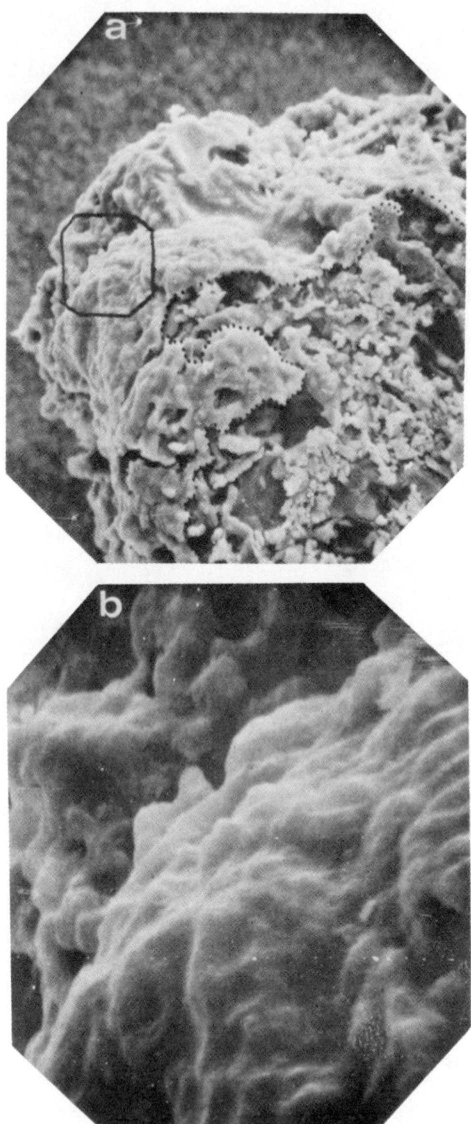

Figure 10.2. *Scanning electron micrograph of 11% w/w $[C_{18}H_{37}SiO_{3/2}]n$ chemically bonded to Chromosorb G (after exhaustive extraction). (a) line of fracture, 750x; (b) outside detail, 3750x. [Reproduced from Aue, W. A., C. R. Hastings, J. M. Augl, M. K. Norr and J. V. Larsen. J. Chromatogr., 56, 295 (1971) with the permission of Elsevier Publishing Company.]*

An excessive activity of silica surface can further result in irreversible adsorption, tailing, and the formation of artifacts during chromatography. While suitable geometrical modifications of silica materials[8] may often cure these deficiencies, these steps are usually accompanied by loss of specific surface area and adsorbent capacity.

Liquid-liquid chromatography can provide the often-needed selectivity of separation, but has its own problems. This method can operate only if the two phases are immiscible, thus giving quite limited perspective of general utility. The immiscible pairs of mobile and stationary phases have vastly different polarities, and "the sample components tend to have partition coefficients in such a system tending either to zero or infinity."[56] Solvent programming techniques cannot be used. Furthermore, even with limited immiscibility of two phases, the stationary phase stripping always takes place to a certain extent, thus resulting in decreased lifetimes of the chromatographic columns and a continuous contamination of recovered fractions. The use of chemically bonded stationary phases (or modified adsorbents) in liquid chromatography is an attractive alternative for the above reasons. Moreover, the systems with unusual selectivities can potentially be prepared through a suitable "tailoring" of siliceous surfaces by the wealth of chemical reactions. Both hydrophobic (reversed-phase) and high polar stationary phases are needed for different applications. The stability of Si-O-Si-C bonds against an electrophilic or nucleophilic attack makes surface silylation particularly attractive and universal for different chromatographic separations.

While the original Martin's work[23] remained relatively unnoticed, Stewart and Perry made a clear point in their publication[56] on the utility of nonpolar packing prepared through the reaction of a diatomaceous earth support with octadecylchlorosilane. Nonpolar silicone phases can now be quite easily prepared and are commercially available from several sources. They are easy to use in the solvent-programming operation; the mobile phases are usually mixtures of water and methanol. Bonded reversed phases have been widely used and different samples were resolved based on this separation principle. Figure 10.3 which shows the gradient elution of model polynuclear aromatic hydrocarbons has been reproduced from the publication of Schmit *et al.*[57] The effect of operating conditions on

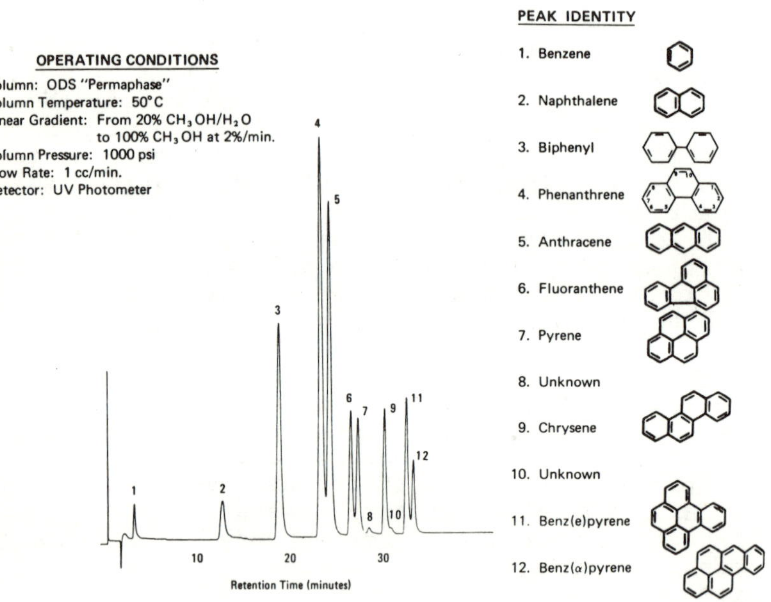

Figure 10.3. Separation of standard polynuclear aromatic hydrocarbons on a reversed-phase bonded material using linear gradient elution. [Reproduced from Schmit, J. A., R. A. Henry, R. C. Williams, and J. F. Dieckman. J. Chromatogr. Sci., 9; 645 (1971) with the permission of Preston Technical Abstracts Company.]

column efficiency and selectivity was studied[57] suggesting benefits of column operation at elevated temperatures.

Relatively little has been reported on the preparation and application of polar silicone-based phases. This situation can be at least partially explained by the limited availability of commercial silane compounds possessing selective groups in the side chain. Although Kiselev[4] pointed out long ago that the silica surfaces can be modified through the reaction of alkoxysilanes containing selective groups, the utilization of this principle in liquid chromatography is relatively recent. Kirkland[49,50] prepared and studied the polar stationary phases with cyano and ether functions bonded chemically to the controlled-surface porosity chromatographic support:

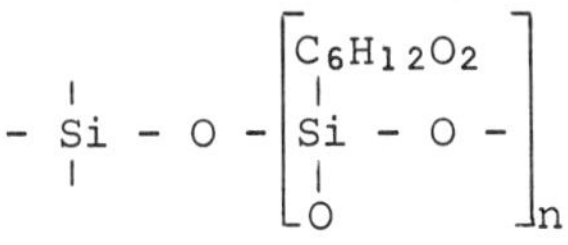

$$-\overset{|}{\underset{|}{Si}}-O-\left[\begin{matrix}C_6H_{12}O_2\\ |\\ Si-O-\\ |\\ O\end{matrix}\right]_n \qquad -\overset{|}{\underset{|}{Si}}-O-\left[\begin{matrix}C_2H_4CN\\ |\\ Si-O-\\ |\\ Y\end{matrix}\right]_n$$

ETHER PHASE 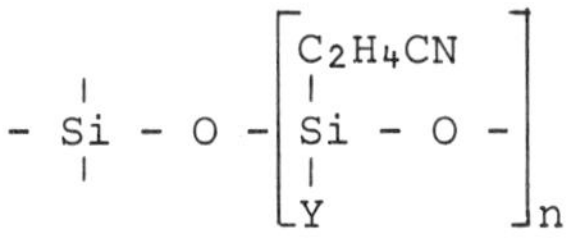 CYANO PHASE

($Y = -CH_3$ or $-O-$)

A different cyano bonded phase has been described by Majors.[58] More recently, Novotny *et al.*[51,52] described the preparation and properties of polar stationary phases prepared by simple modifications of the bonded silicone films prepared through the polymerization of novel silanes specially synthesized for use in the solid-phase peptide synthesis (for review, see reference 6). The chromatographic properties of hydroxy, cyano, and ester phases have been described,[51,52] but many other chemical modifications of the surface structures

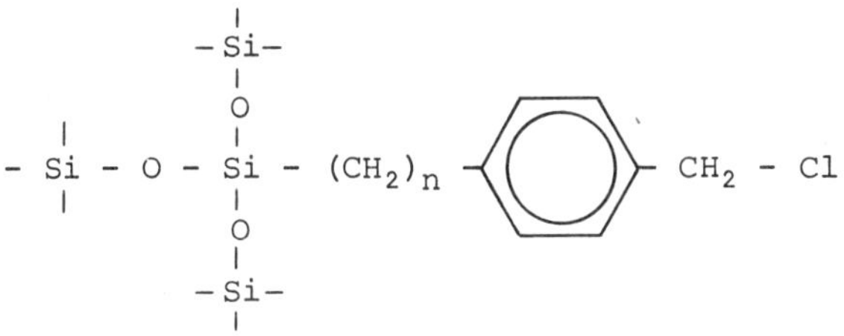

$$-\overset{|}{\underset{|}{Si}}-O-\overset{\overset{\overset{|}{-Si-}}{|}}{\underset{\underset{\underset{|}{-Si-}}{|}}{\overset{O}{\underset{O}{\overset{|}{\underset{|}{Si}}}}}}-(CH_2)_n-C_6H_4-CH_2-Cl$$

are possible.[55] A distinct advantage of this approach is that the stationary phases of different polarities can be essentially all prepared from the identical monomer (or mixture of monomers), thus eliminating thankless tasks of synthetic attempts to provide silanes with the whole scale of polarities and the optimization of reaction conditions for each individual monomer. Since mono-, di-, and trisubstituted silanes of the above structure can now be synthesized,[59] it is possible to obtain surface structures ranging from monolayers (prepared from the monochlorosilane) up to the layers with appreciable thickness. The extent of polymerization and polymer consistency can be controlled by a suitable blending of different monomers which are, consequently, reflected in different column selectivities.[51] A chromatogram of the model mixture of nitroaniline isomers on the hydroxy phase can be seen in Figure 10.4. Other routes leading to the synthesis of

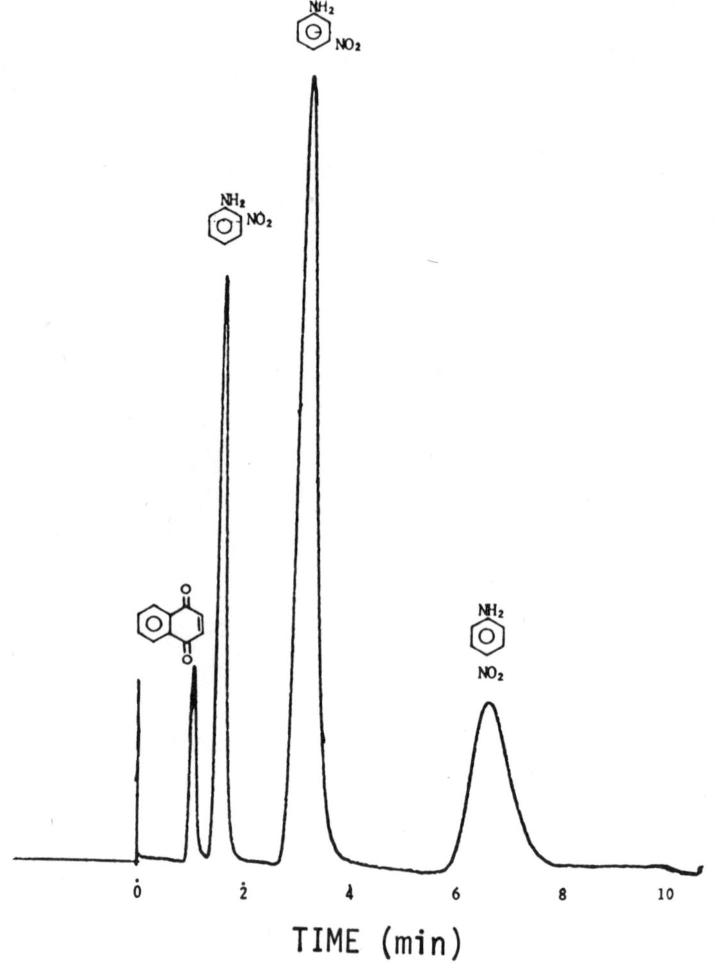

Figure 10.4. *Chromatogram of the model mixture on a hydroxy phase bonded to Corasil I.*
Column: 80 cm x 2 mm, i.d.
Mobile phase: 0.5% isopropanol in heptane
Flow rate: 1.0 ml/min.
[Reproduced from Novotny, M., S. L. Bektesh, K. B. Denson, K. Grohmann, and W. Parr. Anal. Chem., 45, 973 (1973) with the permission of the American Chemical Society.]

polymeric films of glycol and amino structures have also been described.[52] Nitroaniline isomers proved to be useful indicators of column selectivity (polarity) in comparing different bonded phases as seen in Table 10.1, where rather unexpected selectivities with glycol and amino polymers may be noticed. In addition to an effective separation of common polar molecules, some polar phases seem to be useful for chromatography of aromatic systems as suggested by Figure 10.5.

It can be stated that the knowledge of bonded phases used in gas chromatography is more advanced (although still far from complete) than in liquid chromatography. Systematic studies have yet to be done. Many additional phenomena typical for liquid chromatography further complicate the situation. It was found almost invariably[50-52,58] that the chemically bonded phases in liquid chromatography provide somewhat lower column efficiencies than conventional coated substrates. Kirkland[55] first reported that polar silicones performed poorly with nonpolar mobile phases. The solvent penetration into the polymer structure is essential for their proper functioning as chromatographic media. This problem can be overcome by an addition of polar constituent ("moderator") into the mobile phase. Solvated polymer layers then often behave as chromatographic media with unusual selectivities. As suggested by Kirkland,[50] both polymer and solvation agent thus contribute to the solute retention. In general, both column efficiencies and k-values are influenced by the addition of a polar constituent. It has been felt[50-52,58] that the bonded polymers have rather resinous consistency and high viscosity. Partition ratios were found to decrease with increasing flow rates.[50] These phenomena are becoming more critical at higher polymer loadings, thus suggesting limited mass transfer as an explanation. Depending on the nature of both polymer and modifier, the solvation effect is a time-limited process. This situation undoubtedly presents problems for gradient elution techniques. Dependence of k-values on the equilibration time for different bonded phases possessing the same functional groups was measured.[52] Figure 10.6 demonstrates the changes of k-values of p-nitroaniline with the time of solvation for two different polymeric and a monolayer cyano phase (see also Table 10.1). The monolayer phases based on the reaction with a suitable monochlorosilane are more stable analogues of "brushes"[54] and deserve more attention in future studies.

Table 10.1

Average Relative Selectivities of Different Packings of Chloro- and Alkoxysilanes

Monomer	Modification of the Stationary Phase	Functional Groups	k-Value		
			o-Nitroaniline	m-Nitroaniline	p-Nitroaniline
None*	None	Surface silanol groups	0.95	2.74	7.70
Trichloro-[3-(4-chloromethylphenyl)-propyl] silane and dichloromethyl-[4-(4-chloromethylphenyl)-butyl]silane (1:1)	Hydrolysis with a slightly alkaline solution	Hydroxy	0.98	3.01	8.62
Same	Reaction with potassium cyanide	Cyano	1.14	3.12	9.52
Same	Hydrolysis and subsequent esterification of the hydroxy phase	Ester	1.04	2.71	8.40
Monochlorodimethyl-[4-(4-chloromethylphenyl)butyl] silane	Reaction with potassium cyanide	Cyano	1.11	3.30	9.07
Allytrichlorosilane and allylmethyldichlorosilane (1:1)	None	Double bond	1.00	3.16	7.40
Same	Oxidation with H_2O_2 in acetic acid	Glycol (epoxy?)	1.09	5.35	8.95
β-Cyanoethyltriethoxysilane	None	Cyano	0.94	3.32	8.54
N-(β-Aminoethyl)-γ-aminopropyltrimethoxysilane	None	Amino	2.55	6.36	26.60
Allyltriethoxysilane	Oxidation with H_2O_2 in acetic acid	Glycol (epoxy?)	0.83	24.40	9.15

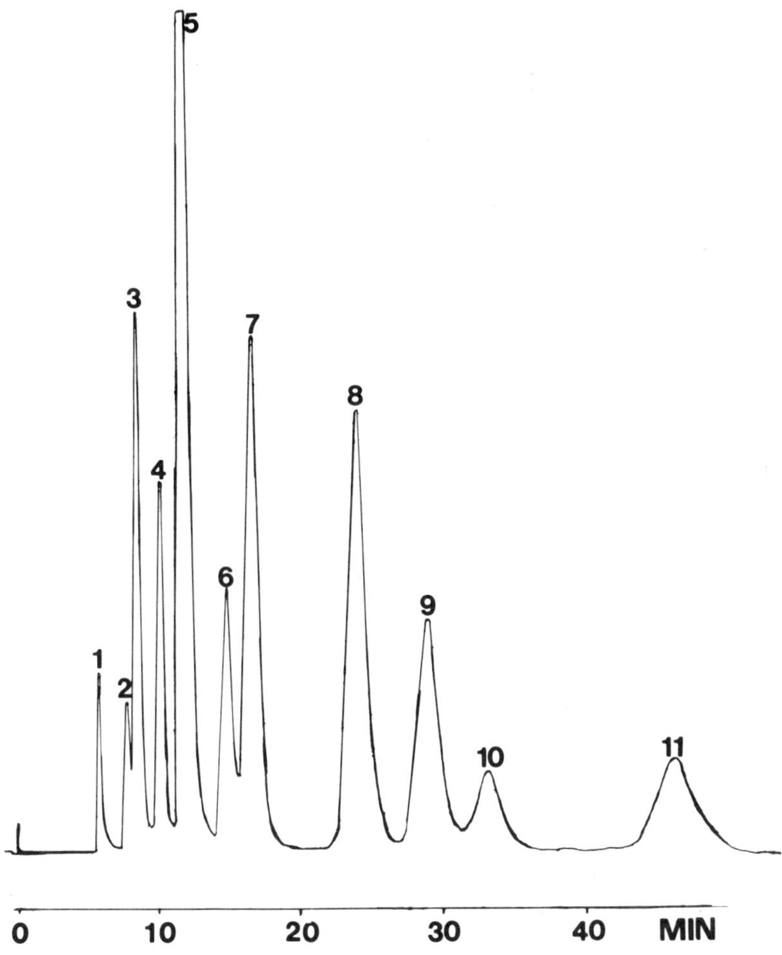

Figure 10.5. Chromatogram of standard polynuclear aromatic hydrocarbons on an amino silane polymer bonded to 10μ silica at room temperature.
Chromatographic conditions: 50 cm x 2.6 mm, i.d. column
Mobile phase: n-hexane, 0.5 ml/min
Peaks 1-11 are: benzene, naphthalene, biphenyl, fluorene, anthracene, pyrene, benzofluorene, triphenylene, benzo[a]pyrene, perylene and dibenzoanthracene.
[Lee, M. L. and M. Novotny, unpublished experiments.]

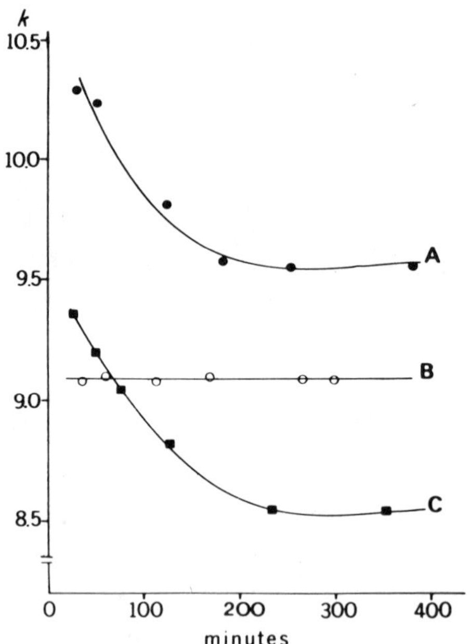

Figure 10.6. Dependence of k-values of p-nitroaniline on the equilibration time for different bonded phases possessing cyano groups.
Conditions: columns (80 cm x 2 mm, i.e.) with cyano phases bonded to Porasil C (35-75 μm)
Flowrate: 1.0 ml/min; room temperature.
Monomers: A=1:1 mixture of trichloro-[3-(4-chloromethylphenyl)propyl]- and dichloromethyl-[4-(4-chloromethylphenyl)butyl]-silane; B=monochlorodimethyl-[4-(4-chloromethylphenyl) butyl]silane; C=β-cyanoethyltriethoxysilane.
[Reproduced from Novotny, M., S. L. Bektesh, and K. Grohmann. J. Chromatogr., 83, 25 (1973) with the permission of Elsevier Publishing Company.]

It is necessary to emphasize that other chromatographic uses of silicone-based polar bonded phases are likely to emerge in the near future. Silicone-based ion-exchangers[55] may well compete with the traditional materials such as polymeric beads or pellicular resins. More importantly, biologically significant molecules may be attached to the solid silica matrices via suitable silane linkages, thus providing extremes of selectivity so urgently needed in biochemical separations. Synthetic efforts heading in this direction have already been initiated.[6,60]

The potential of porous glass and macroporous silica with controlled sizes of pores between 10^2-10^4Å is quite obvious. Ordinary gels used traditionally in biochemical separations suffer from apparent disadvantages such as physical changes under different conditions of pH and ionic strength and, in particular, unsuitability for operation under the conditions of high pressure and flow. However, free silanol groups which may cause structural alterations of sensitive biological molecules must be masked.[61,62] On the other hand, the hydrophilic nature of the separation medium needs to be retained. Eltekov *et al.*[63] have recently demonstrated that human serum albumin, cytochrom c, α-chymotrypsin, and lysozyme can be successfully chromatographed on a macroporous silica (80 m^2/g, with average pore diameter 500Å). Due to the porosity consideration, it is likely that the monolayer approach will be preferable, provided that enough shielding of macromolecules from the silica framework is secured.

Chromatographic Concentration Methods

Determination of trace organics in air and water is of both fundamental scientific and practical importance. Recent concern about ecological problems is directly reflected in an intensive search for improved analytical methods. With the present sensitivity of chromatographic detectors, it usually becomes mandatory to concentrate the concerned compounds before the chromatographic separation and quantitation. The precolumn concentration prior to chromatography is the most common method; both adsorbents and partitioning media have been used. The use of support-bonded silicones for sampling in both air and water pollution was explored by Aue *et al.*[64,65] Heavily-loaded long-chain nonpolar silicones were used to concentrate car exhaust gases and other sources of air pollution.[64] The quantitative aspects of the extraction of chlorinated compounds from different waters were also studied at low concentration levels[65] using the same bonded substrates.[38] Because the bonded phases are nonextractable with organic solvents, small amounts of nonpolar solvent are used to extract organic materials which had accumulated on the sampling cartridge, and the concentrated solution can further be analyzed.

SURFACE MONOLAYERS AND WETTABILITY

Many future improvements in the column technology in gas-liquid chromatography are strongly dependent on our better understanding of wetting phenomena. The effects which take place in the interfacial region between the siliceous solid surface and a coated phase are extremely important for the preparation of inert and efficient chromatographic columns. These considerations apply quite generally for any siliceous material used in chromatography. However, our discussion will be primarily concerned with glass capillary columns which have a unique potential in the separation of extremely complex mixtures, particularly those of biologic origin.[66-68] Although glass capillary columns were used with great success for hydrocarbon analysis more than ten years ago (for review see reference 69), their potential in biochemical separations has not become topical until quite recently. The limited wettability of glass surfaces has been the most serious obstacle to a wider application of glass capillary columns in general. The column technology, properties of glass capillary columns, and many biochemical applications have been reviewed.[67,68]

The surface chemistry of the inner walls of glass capillaries exerts the primary control over the application of thin and homogeneous films of various stationary phases and the necessary column deactivation. It has been common experience that glass in its unmodified form is difficult to wet with organic fluids. This fact had been realized by surface chemists[70,71] long before becoming topical in chromatography. The wettability problems can partially be overcome by decreasing the contact angles of wetting liquids through increased surface area.[67] However, it is also necessary to recognize that the chemical similarities between the wall of a glass capillary and the stationary phase are of great importance in the preparation of homogeneous and stable films.[41,67,72] The recent publications discuss this subject more specifically.[41,42,73]

A more lengthy discussion of the basic modification methods for the surface silanol groups is contained in a review article by Novotny and Zlatkis[67] who also introduced the surface compatibility model. Silane compounds which are likely to form a stable monomolecular layer on the surface of glass are a natural choice. Thus, hydrophobic monolayers are easily formed by a simple trialkylsilylation[4,12] which then exhibit an increased wettability with nonpolar stationary phases, while making the coating of polar films even more difficult. The dependence of film thickness and column efficiency on the chemical treatment had been observed by Novotny and Bartle,[72] but no general method for the preparation of selective monolayers could be suggested until recently.[41] Monochlorodimethyl [4-(4-chloromethylphenyl)butyl]-silane prepared by Grohmann and Parr[59] for the solid-phase peptide synthesis,[6,39,40] proved to provide for a suitable link between the siliceous surface and the stationary phase of a given polarity through the numerous possibilities for chemical modification of the terminal group:[41,51,52]

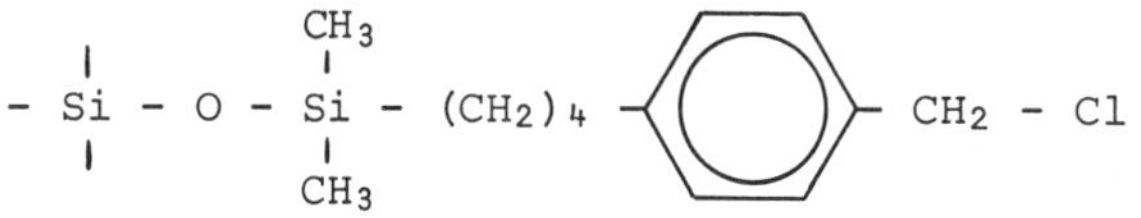

$$-\overset{|}{\underset{|}{Si}}-O-\overset{CH_3}{\underset{CH_3}{Si}}-(CH_2)_4-C_6H_4-CH_2-Cl$$

Thus, hydroxy groups can easily be generated through the hydrolysis with a slightly alkaline solution, cyano groups can be introduced into the surface structures by reaction with potassium cyanide, etc. Many other modifications are possible to match the properties of a given stationary phase (Figure 10.7). This working compatibility model has now been confirmed by more detailed surface studies[42] which further support the importance of chemical similarities between the surface layer and the wetting liquid. Efficient glass capillary columns can now be prepared as based on this approach. Figure 10.8 demonstrates an efficient separation of model methyl esters on a 12-meter column with a modified polyethylene glycol coated on the hydroxy monolayer.

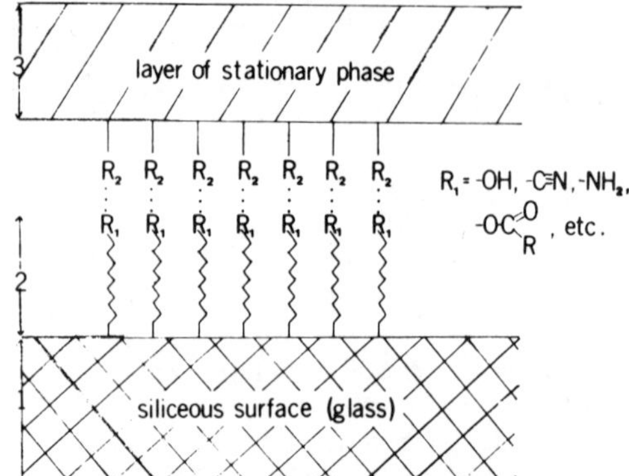

Figure 10.7. Model of chemically compatible surface structures.
1=part of the glass silica framework
2=monolayer possessing a selective group R_1
3=layer of stationary phase possessing R_2, which is chemically compatible with R_1.
[Reproduced from Novotny, M. and K. Grohmann, J. Chromatogr., 84, 167 (1973) with the permission of Elsevier Publishing Company.]

It is very likely that the procedures leading to the preparation of selective monolayers may find additional use outside the capillary column field. The problems of deactivation, in general, are closely associated with both blocking of silanol groups and the wettability phenomena. An extension of this work[41,42] to ordinary chromatographic materials (diatomaceous earths, silica gels, glass beads, etc.) would be quite natural. Moreover, according to Kováts,[74] well defined "tailor-made" surfaces coated with standard pure stationary liquids will be necessary for acquiring more reproducible chromatographic data.

CONCLUSIONS

The most important aspects of the surface silylation pertaining to various chromatographic methods have been reviewed in this chapter. Due to never-ending demands for higher separation efficiencies and sensitive detection in gas and modern liquid chromatography, this area of research is likely to expand further.

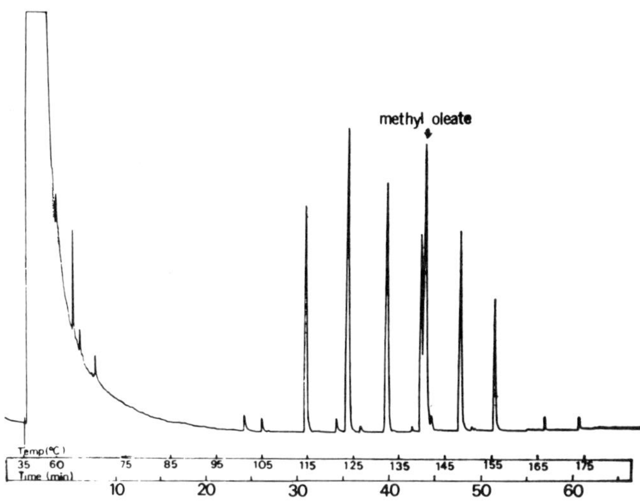

Figure 10.8. Chromatogram of C_{15}-C_{20} fatty acid methyl esters (major constituents) on a glass capillary column provided with a hydroxy monolayer on unetched glass and coated with 3% STAP (a modified Carbowax obtained from Supelco, Bellafonte, Pennsylvania). Column: 12m x 0.3 mm, i.d. Sample injected without splitting. [Reproduced from Novotny, M. and K. Grohmann, J. Chromatogr., 84, 167 (1973) with the permission of Elsevier Publishing Company.]

With regard to effective deactivations and wettability problems, more attention will have to be paid to the controlled preparation of well-defined siliceous surfaces as well as the synthesis and modifications of suitable silanes. At the present stage of development, it appears that chemically bonded silicone phases will find more practical use in liquid rather than gas chromatography. Increased activity is particularly anticipated in the field of macromolecular separations. Chemically bonded silicones in liquid chromatography will have to be studied more thoroughly in order to understand their interactions with chromatographed solutes. More systematic use of both purely chromatographic and other physicochemical methods is necessary.

ACKNOWLEDGMENT

Valuable comments of Dr. Walter A. Aue concerning this chapter are greatly appreciated. This work was supported by the National Science Foundation Grant No. GP-33751.

REFERENCES

1. Phillips, C. S. G., M. J. Walker, C. R. McIlwrick, and P. A. Rosser. "Slow-Processes and Gas Chromatography," *J. Chromatogr. Sci.*, *8*, 401 (1970).
2. Kiselev, A. V. "Adsorbents in Gas Chromatography," *Adv. Chromatogr.*, *113* (1967).
3. Kiselev, A. V. and Y. I. Yashin. *Gas-Adsorption Chromatography.* (New York: Plenum Press, 1969).
4. Kiselev, A. V. "The Importance of the Solid Surface in Partition, Adsorption and Capillary Gas Chromatography" (Third Opening Lecture) in *Gas Chromatography 1962*, M. van Swaay (Ed.) (London: Butterworths, 1962).
5. Blackman, L. C. F. and R. Harrop. "Hydrophilation of Glass Surfaces. II. Reaction of Quaternary Ammonium Compounds with Surface Hydroxyl Groups," *J. Appl. Chem.*, *18*, 43 (1968).
6. Parr, W. and M. Novotny. "Use of Novel Silanes for the Solid-Phase Peptide Synthesis and the Preparation of Polar Chemically Bonded Phases for Liquid Chromatography," Chapter 10 this book.
7. Ottenstein, D. M. "The Chromatographic Support," *Adv. Chromatogr.*, *3*, 137 (1966).
8. Bebris, N. K., A. V. Kiselev, B. Y. Mokeev, Y. S. Nikitin, Y. I. Yashin, and G. E. Zaizeva. "Macroporous Silica as Adsorbent for Molecular Chromatography," *Chromatographia*, *4*, 93 (1971).
9. Zhdanov, S. P., V. I. Kalmanovskii, A. V. Kiselev, M. M. Fiks and Y. I. Yashin. "Use of Porous Glasses as Adsorbents in Gas Chromatography," *Russ. J. Phys. Chem.*, *36*, 595 (1962).
10. Mohnke, M. and W. Saffert. "Adsorption Chromatography of Hydrogen Isotopes with Capillary Columns," in *Gas Chromatography 1962*, M. van Swaay (Ed.) (London: Butterworths, 1962).
11. Bruner, F. A. and G. P. Cartoni. "Use of Glass Capillary Columns with Modified Internal Area in Gas Chromatography," *Anal. Chem.*, *36*, 1522 (1964).
12. Novotny, M. and K. Tesarik. "Surface Treatment of Glass Capillaries for Gas Chromatography," *Chromatographia*, *1*, 332 (1968).

13. Iler, R. K. *The Colloid Chemistry of Silicas and Silicates.* (Ithaca, New York: Cornell University Press, 1955).
14. Davydov, V. Y., L. T. Zhuravlev, and A. V. Kiselev. "Infrared Spectra Due to Surface and Bulk Hydroxyls and Silica," *Russ. J. Phys. Chem., 37,* 1404 (1963).
15. Armistead, C. G., A. J. Tyler, F. H. Hambleton, S. A. Mitchell, and J. A. Hockey. "The Surface Hydroxylation of Silica," *J. Phys. Chem., 73,* 3947 (1969).
16. Unger, K. "Structure of Porous Adsorbents," *Angew. Chem. Intern. Ed. Engl., 11,* 267 (1972).
17. Gilpin, R. K. and M. F. Burke. "Role of Tri- and Dimethylsilanes in Tailoring Chromatographic Adsorbents," *Anal. Chem., 45,* 1383 (1973).
18. Snyder, L. R. *Principles of Adsorption Chromatography.* (New York: M. Dekker, 1968).
19. Kiselev, A. V. "Nonspecific and Specific Interactions of Molecules of Different Electronic Structures with Solid Surfaces," *Discussions Faraday Soc., 40,* 205 (1965).
20. Kiselev, A. V. and V. I. Lygin. "Infrared Spectroscopy of Solid Surfaces and Adsorbed Molecules," *Surface Sci., 2,* 236 (1964).
21. Basila, M. R. "Hydrogen Bonding Interaction Between Adsorbate Molecules and Surface Hydroxyl Groups on Silica," *J. Chem. Phys., 35,* 1151 (1961).
22. Gavrilova, T. V., M. Krejci, H. Dubsky, and J. Janak. "Bestimmung der Thermischen Stabilität von Chemisch Modifizierenden Schichten an der Adsorbenten- und Trägeroberfläche in der Gaschromatographie," *Collection Czechosl. Chem. Commun., 29,* 2753 (1964).
23. Howard, G. A. and A. J. P. Martin. "The Separation of C_{12}-C_{18} Fatty Acids by Reversed-Phase Partition Chromatography," *Biochem. J., 56,* 532 (1950).
24. Unger, K., K. Berg, and E. Gallei. "Herstellung Oberflächen-Modifizierter Adsorbentien," *Kolloid-Z.Z. Polymer, 234,* 1108 (1969).
25. Sorell, J. B. and R. Rowan, Jr. "Nitrogen Adsorption on Chemically Modified Silica Gels," *Anal. Chem., 42,* 1712 (1970).
26. Borisenko, I. V., A. V. Kiselev, R. S. Petrova, V. K. Chuikina, and K. D. Shcherbakova. "Chemical Modification of the Surface of Silica Gel by Methylchlorosilanes for Gas Chromatography," *Russ. J. Phys. Chem., 39,* 1436 (1965).
27. Bens, E. M. and C. M. Drew. "Diatomaceous Earth: Scanning Electron Microscope of 'Chromosorb P,'" *Nature, 216,* 1046 (1967).
28. DeMets, M. and A. Lagasse. "Scanning Electron Micrographs of Column Packing," *J. Chromatogr. Sci., 8,* 272 (1970).

29. Aue, W. A., C. R. Hastings, J. M. Angl, M. K. Norr, and J. V. Larsen. "The Distribution of Support-Bonded Silicones on Chromosorbs," *J. Chromatogr., 56,* 295 (1971).
30. Giddings, J. C. "Liquid Distribution on Gas Chromatographic Support," *Anal. Chem., 34,* 458 (1962).
31. Perrett, R. H. and J. H. Purnell. "A Study of the Reaction of Hexamethyldisilazane with Some Common Gas-Liquid Chromatographic Solid Supports and Its Effect on Their Adsorptive Properties," *J. Chromatogr., 7,* 455 (1962).
32. Horning, E. C., E. Moscatelli, and C. C. Sweeley. "Polyester Liquid Phases in Gas-Liquid Chromatography," *Chem. Ind. (London), 751,* (1959).
33. Bohemen, J., S. H. Langer, R. H. Perrett, and J. H. Purnell. "A Study of the Adsorptive Properties of Firebrick in Relation to its Use as a Solid Support in Gas-Liquid Chromatography," *J. Chem. Soc., 2444* (1960).
34. Dewar, R. A. and V. E. Maier. "High-Efficiency Glass Bead Columns for Gas Chromatography," *J. Chromatogr., 11,* 295 (1963).
35. Urone, P. and J. F. Parcher. "Deactivation of Gas Chromatographic Support by Radiation-Induced Copolymerization," *J. Gas Chromatogr., 3,* 34 (1965).
36. Holmes, W. L. and E. Stack. "Gas Chromatography of Squalene, Sterols and Bile Acid Methyl Esters," *Biophys. Acta, 56,* 163 (1962).
37. Kabot, F. J. and L. S. Ettre. "Instrumental Aspects of Steroid Analysis by Gas Chromatography," *J. Gas Chromatogr., 2,* 21 (1964).
38. Aue, W. A. and C. R. Hastings. "Preparation and Chromatographic Uses of Surface-Bonded Silicones," *J. Chromatogr. 42,* 319 (1969).
39. Parr, W. and K. Grohmann. "A New Solid Support for Peptide Synthesis," *Tetrahedron Lett., 28,* 2633 (1971).
40. Parr, W. and K. Grohmann. "Solid Phase Peptide Synthesis on an Inorganic Matrix Having Organic Groups on the Surface," *Angew. Chem. Intern. Ed. Engl., 11,* 314 (1972).
41. Novotny, M. and K. Grohmann. "Selective Monomolecular Layers for Improved Wettability of Glass Capillary Columns for Gas-Liquid Chromatography," *J. Chromatogr., 84,* 167 (1973).
42. Bartle, K. D. and M. Novotny. "Correlations of Surface Measurements with Chromatographic Performance of Chemically-Modified Glass Capillary Columns," submitted for publication.
43. Vinogradov, G. V., L. V. Tikova, N. V. Akshinskaya, N. K. Bebris, A. V. Kiselev, and Y. S. Nikitin. "Aerogels of Organic and Organosilicon Polymers with a High Surface Area," *Russ. J. Phys. Chem., 40,* 472 (1966).
44. Hastings, C. R., W. A. Aue, and J. M. Angl. "Surface-Bonded Silicones from Volatile Monomers for Chromatography," *J. Chromatogr., 53,* 487 (1970).

45. Hastings, C. R., W. A. Aue and F. N. Larsen. "*In Situ* Synthesis of Silicone Liquid Phases for Chromatography," *J. Chromatogr.*, *60*, 329 (1971).
46. Abel, E. W., F. H. Pollard, P. C. Uden, and G. Nickless. "A New Gas-Liquid Chromatographic Phase," *J. Chromatogr.*, *22*, 23 (1966).
47. Bossart, C. J. "Chemically Bonded Liquid-Phase Packings for Gas Chromatography," *ISA Trans.*, *7*, 283 (1968).
48. Al-Taiar, A. H., J. R. L. Smith, and D. J. Waddington. "Preparation of Thermally Stable Gas-Chromatographic Packing Materials," *Anal. Chem.*, *42*, 935 (1970).
49. Kirkland, J. J. and J. J. DeStefano. "Controlled Surface Porosity Support with Chemically-Bonded Organic Stationary Phases for Gas and Liquid Chromatography," *J. Chromatogr. Sci.*, *8*, 309 (1970).
50. Kirkland, J. J. "High-Speed Liquid Partition Chromatography with Chemically Bonded Organic Stationary Phases," *J. Chromatogr. Sci.*, *9*, 206 (1971).
51. Novotny, M., S. L. Bektesh, K. B. Denson, K. Grohmann, and W. Parr. "Polar Silicone-Based Chemically Bonded Stationary Phases for Liquid Chromatography," *Anal. Chem.*, *45*, 971 (1973).
52. Novotny, M., S. L. Bektesh, and K. Grohmann. "Chemically Bonded Stationary Phases with Variable Selectivity," *J. Chromatogr.*, *83*, 25 (1973).
53. Bossart, C. J. "Gas-Chromatographic Column and Method of Making Same," U.S. Patent No. 3,514,925 (1970).
54. Halász, I. and I. Sebastian. "New Stationary Phases for Chromatography," *Angew. Chem. Intern. Ed. Engl.*, *8*, 453 (1969).
55. Novotny, M. unpublished experiments.
56. Stewart, H. N. M. and S. G. Perry. "A New Approach to Liquid Chromatography," *J. Chromatogr.*, *37*, 97 (1968).
57. Schmit, J. A., R. A. Henry, R. C. Williams, and J. F. Dieckman. "Applications of High Speed Reversed-Phase Liquid Chromatography," *J. Chromatogr. Sci.*, *9*, 645 (1971).
58. Majors, R. E. "Techniques for Liquid Chromatographic Columns Packed with Small Porous Particles," *Anal. Chem.*, *45*, 4 (1973).
59. Part, W., K. Grohmann, and K. Haegele. "Silikon-Matrices für die Festphasen-Peptidsynthese," *Annal. Chem.*, in press.
60. Grushka, E. and R. P. W. Scott. "Polypeptide as a Permanently Bound Stationary Phase in Liquid Chromatography," *Anal. Chem.*, *45*, 1626 (1973).
61. Johnson, P. and T. L. Whateley. "The Effect of Glass and Silica Surfaces on Trypsin and α-chymotrypsin Kinetics," *Biochim. Biophys.*, *276*, 323 (1972).
62. Hawk, G. C., J. A. Cameron, and L. B. Dufault. "Chromatography of Biological Materials on Polyethylene

Glycol Treated Controlled-Pore Glass," *Prep. Biochem.*, *2*, 193 (1972).

63. Eltekov, Y. A., A. V. Kiselev, T. D. Khokhlova, and Y. S. Nikitin. "Adsorption and Chromatography of Proteins on Chemically Modified Macroporous Silica-Aminosilochrom," *Chromatographia*, *6*, 187 (1973).
64. Aue, W. A. and P. M. Teli. "Sampling of Air Pollutants with Support-Bonded Chromatographic Phases," *J. Chromatogr.*, *62*, 15 (1971).
65. Aue, W. A., S. Kapila, and C. R. Hastings. "The Use of Support-Bonded Silicones for the Extraction of Organochlorines of Interest from Water," *J. Chromatogr.*, *73*, 99 (1972).
66. Novotny, M. and A. Zlatkis. "High-Resolution Chromatographic Separation of Steroids with Open Tubular Glass Columns," *J. Chromatogr. Sci.*, *8*, 346 (1970).
67. Novotny, M. and A. Zlatkis. "Glass Capillary Columns and Their Significance in Biochemical Research," *Chromatogr. Rev.*, *14*, 1 (1971).
68. Novotny, M. "Glass Capillary Columns: Their Applicability to Biochemical and Health-Related Problems," in *Chromatography 1962*, S. G. Perry (Ed.) (Applied Science Publishers, 1973), p. 25.
69. Desty, D. H. "Capillary Columns: Trials, Tribulations and Triumphs," *Adv. Chromatogr.*, *1*, 199 (1965).
70. Fox, H. W., E. F. Hare, and W. A. Zisman. "The Spreading of Liquids on Low-Energy Surfaces. VI. Branched-Chain Monolayers, Aromatic Surfaces and Thin Liquid Films," *J. Colloid Sci.*, *8*, 194 (1953).
71. Fox, H. W., E. F. Hare, and W. A. Zisman. "Wetting Properties of Organic Liquids on High-Energy Surfaces," *J. Phys. Chem.*, *59*, 1097 (1955).
72. Novotny, M. and K. D. Bartle. "Surface Modification of Glass Capillaries for Gas Chromatography by Silylation," *Chromatographia*, *3*, 272 (1970).
73. Bartle, K. D. "Thickness of the Stationary-Phase Film in Glass Open Tubular Columns for Gas Chromatography Coated by the Dynamic Method," *Anal. Chem.*, *45*, 1831 (1973).
74. Kováts, sz. E. personal communication.

INDEX

UNIV. COLL. N.W.
BANGOR
LIBRARY